P9-DFT-898

Also by Dick Teresi

Lost Discoveries

Laser *(with Jeff Hecht)*

The God Particle *(with Leon Lederman)*

The 3-Pound Universe *(with Judith Hooper)*

The Undead

The Undead

ORGAN HARVESTING, THE ICE-WATER TEST,

BEATING-HEART CADAVERS—

HOW MEDICINE IS BLURRING THE LINE

BETWEEN LIFE AND DEATH

DICK TERESI

Pantheon Books • New York

Library of Congress Cataloging-in-Publication Data
Teresi, Dick.
The undead : organ harvesting, the ice-water test, beating-heart cadavers—
how medicine is blurring the line between life and death / Dick Teresi.
p. cm.
Includes bibliographical references and index.
ISBN 978-0-375-42371-0
Title.
[DNLM: 1. Death. 2. Attitude to Death. 3. Death—Autobiography.
4. Persistent Vegetative State. 5. Tissue and Organ Harvesting. W 820]
LC classification not assigned 610—dc23 2011032025

www.pantheonbooks.com

Jacket photograph by Nicholas Eveleigh/Photodisc/Getty Images
Jacket design by Peter Mendelsund

Printed in the United States of America

First Edition

2 4 6 8 9 7 5 3 1

Contents

Acknowledgments

Thanks to Judith Hooper and Janet MacFadyen for their extensive research, and to Shelley Wanger, Lynn Nesbit, Ken Schneider, and Juhea Kim for their efforts and support. Lynn Anderson provided deft copyediting.

Prologue

"B" POD of the intensive care unit (ICU) at Baystate Medical Center is very bright: intense lights, everything painted in primary colors. Cabinets, counters, and chairs are a sparkling blue, obviously intended to be cheery, not sepulchral. A blue central stripe on the grayish linoleum floor leads us from room to room. There are three ICU pods at this Springfield, Massachusetts, hospital, each with three rooms arranged like spokes around a central station filled with computer terminals. Doctors and nurses come and go.

The patients in their separate rooms, like sickly fish in glass tanks arranged for observation, are largely slack-mouthed and gray or pale yellow in color. Most are fitted with clear plastic tubes the diameter of vacuum-cleaner hoses attached to their mouths. Most are old and gnarled, with sallow, sickly complexions. Sometimes one can see a bare concave chest or a yellowed foot sticking out from beneath a sheet. None is conversing.

We are not here to see them. We have come for the star of "B" Pod. We will call her Fernanda, a fifty-seven-year-old woman of Portuguese descent, who looks as though she might have just come back from a day at the mall. Her dark olive skin appears healthy against the sheets. Her features do not show the ravages of suffering or pain. Her eyes are closed beneath the heavy arched eyebrows. She has thick black eyelashes. Her wiry black hair with strands of gray is arranged neatly on the pillow. She has delicate legs and exquisite feet. Her face is serene; her chest rises and falls with the familiar rhythm of normal human breathing. She looks as if she had been put to sleep by a wicked queen in a fairy tale and needs only to be reanimated. Her vital signs, displayed on a monitor in squiggles the bright green of coloring books, are normal. She looks better than I do.

We are all there, including a group of medical students and interns, to watch Dr. Thomas Higgins pronounce Fernanda brain dead. She had been at work when coworkers found her on the floor and threw water on her face, a tactic that doesn't work well against ischemic stroke. She probably had a brief headache, says Higgins, then fell unconscious. The ICU doctors at Baystate feel she will not recover. The brain-death team is about to make it official.

Those expecting space-age equipment, sophisticated brain scans, and the like, won't find it. The exam is conducted mostly with tools you could find around your home: a flashlight, a Q-tip, some ice water. Her reflexes are tested. A light is shined in her eyes. Her head is turned from side to side. Cold water is squirted into an ear. She is disconnected from her ventilator to make sure she can't breathe on her own. In less time than my ophthalmologist took to prescribe my last pair of bifocals, Fernanda is declared brain dead. That means she is legally dead, just as dead under the law as if her heart had stopped beating. Fernanda is then hooked back up to her ventilator to keep her organs fresh for transplant. Her heart continues to beat; her lungs continue to breathe. Though dead, she remains the best-looking patient in the ICU. The declaration of her death was more philosophical than physiological. A nurse says, "Whatever it was that made her *her* isn't there anymore."

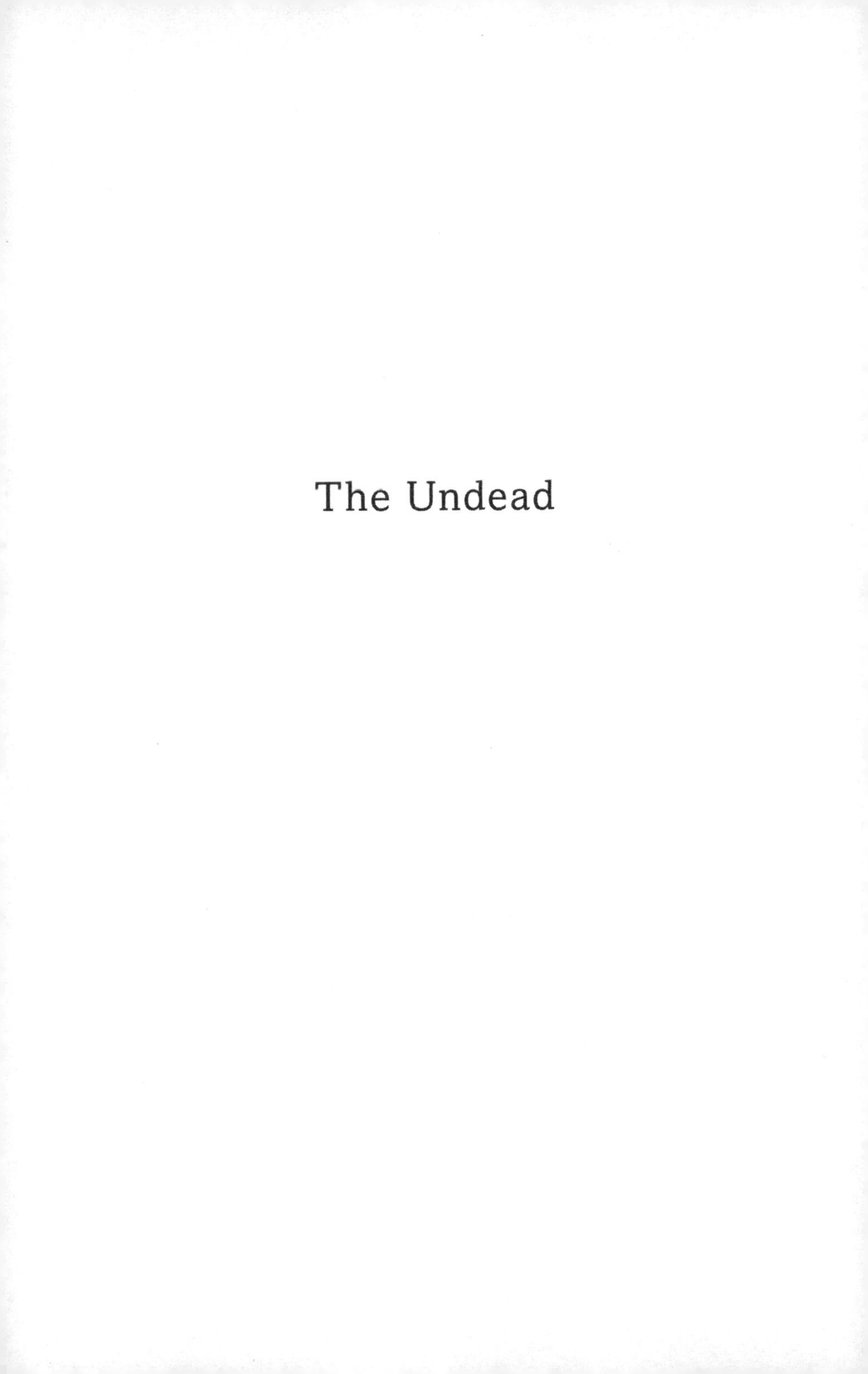

The Undead

ONE

Death Is Here to Stay

For all the accomplishments of molecular biology, we still can't tell a live cat from a dead cat.

—Lynn Margulis

ARE YOU dead or alive? A dumb question, it would seem. If you're reading this book, you are most likely alive. You know it, but do those in control know it? Will they acknowledge it? These are no longer stupid questions. The bar for being dead has been lowered. The bar for being considered alive has been raised. The old standards for life—Are you breathing? Is your heart beating? Are your cells still intact, not putrifying?—have been abandoned by the medical community in favor of a more demanding standard. Are you a person? Is what makes you *you* still intact? Can you prove it? Such concepts were previously the domain of philosophers and priests, but today it is doctors who determine our legal humanity. The dead are also not immune from judgment. A presidential council on bioethics recently determined that some dead people are less "healthy" than others. It is a different world.[1]

This is a book about physical death. It began as a simple magazine article more than a decade ago, a report on the state of the art of death determination. I assumed I would find high-tech medical equipment and techniques that would tell us when a human being had stopped living, that would pinpoint the moment that "what made her *her*" was gone. I eventually abandoned this goal and the article itself. Humans have long lived in denial about their own deaths, but I discovered that this denial has spread to the medical establishment, even to our beliefs about who is dead and who is alive. Our technology has not illuminated death; it has only expanded the breadth of our ignorance. Technology indicates that many of our assumptions about life and death, consciousness and unconsciousness, are wrong. Technology is telling us a great deal about our ignorance, but we are ignoring the information. My focus is scientific information about physiological death, but science and cultural factors often compete in an unproductive manner, canceling each other out. Though we have made technological advances, they often remain unused when it comes to dealing with the dying and dead, so cultural factors—philosophy, ethics, economics, religion—cannot be ignored.

In the following chapters, you will meet brain-death experts, undertakers, cell biologists, coma specialists (and those who have recovered from coma), organ transplant surgeons and organ procurers, anesthesiologists who study pain in legally dead patients, doctors who have saved live patients from organ harvests, experimenters who have removed the hearts of dead people and restarted them many hours later, paramedics, doctors who have used brain-dead pregnant women as human incubators, pediatric neurologists who have kept the heart of a legally dead boy pumping for twenty years, nurses who care for beating-heart cadavers, ICU doctors who are now subtly pressured into declaring their patients dead rather than saving them, hospice workers, execution experts, veterinarians who put animals "to sleep," Egyptologists specializing in mummies, doctors who communicate with those in coma and persistent vegetative state (PVS), people who have been frozen alive, doctors who take MRIs of cats while they kill them, doctors who have drained the blood from patients' heads in the course of brain surgery, lawyers, historians of death, theologians, ethicists, and many others.

Death, for most people, is not a comforting topic, and thus in the great mass of nonfiction literature devoted to the topic, death is treated as something that happens to someone else. My temptation is to write this as if I were narrating a dirigible explosion ("Oh, the humanity!"). Someone far away, perhaps in New Jersey, is dying. You, the audience, and I, the announcer, are merely witnesses. Let us reject that fiction. You, the reader, will die. If it is any consolation, keep in mind that I will also die. At my age, sooner rather than later.

During my research, I have spoken about death before groups of people on several occasions: to classes of college students, to groups of senior citizens, to people at dinner parties and other social gatherings. For the most part, those discussions have been disastrous. When I was talking at a dinner about the vagaries of brain death and the fact that our technology cannot ascertain the condition of most of the brain, one woman, a medical doctor, actually rose from her seat and yelled at me.

She said that writing about this topic was "irresponsible," that it would set organ donation back decades. She threatened to "call your editor." In an undergraduate honors class at the University of Massachusetts, a senior premed major became angry with me when I spoke about patients in persistent vegetative state who show signs of consciousness. Her grandmother was in a comatose state, and her family was confused about what to do. The woman's estate was dwindling because of her care, and, I gathered, so were the student's hopes of paying for medical school. The premed student said that her grandmother was no longer "useful." Those were two remarkable cases, but in general I made people uneasy, even angry. They defended their traditional ideas of life and death to me passionately, forcefully. My protestations that I was merely a journalist reporting facts as I found them, not making moral judgments, was of no consolation. I told the angry doctor that she could yank as many organs as she pleased out of people, and I would not stop or condemn her. I told the student that she and her family could pull the plug on Grandmother and I would not say a word. This just made them angrier. Not everyone was upset with the facts I presented. But those who were, were livid.

It was years before I figured out my apparent mistake. I assumed the information I was presenting, which threatens traditional views of death, was upsetting them. I now believe that it was something simpler: I was reminding people that they were going to die. Not someone else. Them.

In 1973, the cultural anthropologist Ernest Becker put forth an unusual thesis. In *The Denial of Death,* his Pulitzer Prize–winning book, Becker said that humans are like other animals, with an evolutionary drive to survive. "Live! Live! Live!" our genes are screaming at us.[2] Unlike other animals,[3] however, said Becker, we humans know we cannot ultimately survive, that we will die. This dilemma, he believed, drives us mad. The awareness of our inevitable annihilation combined with our evolutionary program for self-preservation holds the potential for evoking paralyzing terror. Becker stated, "The result was the emergence of man as we know him: a hyperanxious animal who constantly invents reasons for anxiety even where there are none."[4] Becker wrote that man is terrified of death, and deals with this terror by denying death and keeping it unconscious.[5] He felt that this terror directs a "substantial portion of human behavior," according to one researcher.[6]

Becker's hypothesis was wide-sweeping. Death terror, he said, was the primary reason that humans created culture. Our religions, our political systems, our art, music, and literature—all this we have constructed "to assure ourselves that we have achieved something of lasting worth."[7] Becker said that "everything that man does in his symbolic world is an attempt to deny and overcome his grotesque fate. He literally drives himself into a blind obliviousness with social games, psychological tricks, personal preoccupations so far removed from the reality of his situation that they are forms of madness—agreed madness, shared madness, disguised and dignified madness, but madness all the same."[8] Those who delude themselves into believing they have achieved something of lasting worth would include, I assume, people who write books.

Becker's theory was intriguing, plausible, and explained much of human behavior. Its only flaw was that Becker had no concrete evidence. As a matter of fact, how would one even go about testing the hypothesis? Terror management was the superstring theory of anthropology—fascinating but not testable.

Help came long after Becker had resolved, by dying, his own death terror. In the late 1970s, three graduate students on an intramural bowling team at the University of Kansas began discussing death terror while the pins were being reset. Through the years, Jeff Greenberg, Tom Pyszczynski, and Sheldon Solomon translated Becker's ideas into a formal theory that could be examined empirically. Their work culminated in a series of remarkable "mortality salience" experiments detailed in a 1997 paper.

Subjects were not told the true purpose of the experiment. Irrelevant questions and reading passages were included to mislead them. Embedded in an opening questionnaire, however, were these directions: "Please describe the emotions the thought of your own death arouses in you. Jot down, as specifically as you can, what you think will happen to you physically as you die and once you are physically dead."

Greenberg, Pyszczynski, and Solomon used this and other techniques in different experiments, such as asking subjects to write their own obituaries or flashing the word "death" intermittently on a computer screen for twenty-eight milliseconds. In another case, the researchers used no verbal signals. They simply interviewed subjects in front of a funeral home.[9] Control groups received no such death signals.

At the heart of the experiment were essays, pro-American and anti-American, supposedly written by foreign students studying in the United States but actually written by the psychologists. The pro essays stated that the United States was the greatest country in the world, the land of opportunity and freedom, and so on. The anti essays stated that American ideals were phony and the rich were getting richer, the poor poorer. Those subjects who were subjected to "mortality salience" ranked the pro essayist as extremely likable, and the anti essayist as extremely unlikable. The control group was not nearly so adamant. Greenberg et al. say that those who are made aware of their mortality need to offset their inner terror by defending their worldview and will praise to extremes those who hold the same worldview and denigrate those who hold different values.

The subjects were not consciously in terror. Corpses were not being embalmed before their eyes. The reminders of their mortality, as mentioned, had come earlier in the experiment. The terror over their mortality simmered unnamed in their unconscious, and they circled the wagons, philosophically speaking, scorning those who opposed their worldview and embracing those who agreed.

In the most remarkable of this series of experiments, Tucson, Arizona, judges were given legal briefs of arrests for prostitution and were asked what bail they would set. Half had been given the mortality salience treatment ("Describe the emotions that your own death arouses . . ."), and half had not. The judges who had been reminded of their mortality set bails that were *nine times as high.* Sheldon Solomon said that no one believed they could influence professional judges to stray from their bail guidelines by reminding them of their deaths. Even Solomon was surprised at the ninefold difference between death-reminded judges and controls.[10]

The prostitutes awaiting trial represented threats to the judges' worldview, and they had to be severely punished and denigrated. The original 1997 experiments have been replicated by more than three hundred other experiments in thirteen countries on five continents. When a person's unconscious is roiling with death terror, he will strike out at anyone he considers too different from himself, just as he will embrace his own kind more. Christian participants liked fellow Christians more and Jewish people less. Germans sat farther away from Turks and closer to fellow Germans. Solomon says that our cultural symbols are not strong enough to overcome death, so when our mortality salience is aroused, we "typically respond to people with different beliefs, or scapegoats, by berating them, trying to convert them to our system of beliefs, and/or just killing them."[11]

Solomon admits that Becker was "dismissed as an irrelevant clown by academics," despite the Pulitzer Prize, and that he went over the line by claiming death denial was "the end-all and be-all of a culture." Greenberg, Pyszczynski, and Solomon also had difficulty getting their work published. It was considered the terrified musings of a morose bowling team. There was little support from fellow professors. As Solomon pointed out, "Scientists said our work was ridiculous. English professors said it was obvious." However, the 1997 experiments transformed scientific opinion, especially after other researchers replicated them.[12]

Greenberg, Pyszczynski, and Solomon gained more attention when they applied their experimental method to voting preferences prior to the 2004 U.S. presidential election. Subjects reminded of their mortality preferred "charismatic" leaders over "relationship-oriented" leaders, who have an egalitarian approach, or "task-oriented" leaders, who simply want to get the job done. A charismatic leader was defined as "a supremely confident leader who has a grand vision and can provide self-worth through identification with the leader and the leader's vision."[13] Charismatic leaders don't offer competence or participatory democracy but rather self-esteem. The study was controversial because of the not-very-well-concealed allusions to President George W. Bush, a classic value-driven leader with a mission of defeating evil. With some subjects, the team flashed subliminal 9/11 images. According to this experiment, voters turned out for George Bush not just because he made them feel safe but because he offered the illusion of immortality. He bolstered Americans' self-worth and denigrated those with different religions or worldviews. Similarly, when such studies were conducted in the Middle East, experimenters found that mortality salience prompted Islamic subjects to think more highly of suicide bombers and to consider becoming bombers themselves.[14]

Solomon, who was precociously morbid, became interested in death at nine years old when his grandmother died. The finality of death struck him. "I was not overenthusiastic about death," he says. As an adult, he has taken his research seriously, personally, and quit his job as a Skidmore College professor for a year to work in restaurants and in construction to think out the problem of mortality. "It brought the message home to me. Don't squander a moment." He decided not to squander his PhD, realized that being a professor paid better than washing dishes, and returned to Skidmore.

Why, you may be asking yourself, given this research, am I reminding the reader of his inevitable death? Am I attempting to make superpatriots of you all? Or to inspire foreign readers to strap on a vest of improvised explosive devices (IEDs) and set themselves off in a Tel Aviv mall? Solomon says that it is the *subliminal* awareness of mortality, not conscious awareness, that drove his subjects mad. Drawing on Freud's dictum that making the unconscious conscious allows people to talk about what's bothering them, Solomon says, "When you remind people of their death, and you keep that thought explicit, you have a small window in which they let go of their defenses." He compared the technique to Franciscan monks who surround themselves with human skulls, to keep death foremost in their minds.

When I spoke to college classes, at Solomon's suggestion I asked students to visualize their own deaths using language similar to that in Solomon et al.'s experiments. ("Please describe the emotions the thought of your own death arouses in you. Jot down . . .") Instead of leaving these thoughts to linger unconsciously, I then told them about the mortality salience studies. The classes were no longer hostile, and students were attentive to what I had to say.

Thus, I may have to shake a skull in your face from time to time. For me, the constant exposure to death has lowered my normally high rate of anxiety. My friends say I've mellowed. They ask me why. I tell them, cheerfully, that I'm going to die.

DEAD CATS

On a late November morning, at 6:30 a.m., I found our tuxedo cat, Blitz, on the side of the road. He was stiff to the touch, rigor mortis having set in. Still, he looked like himself in life, the forepaws stretched out in front of him, delicately crossed, the hind legs elegantly extended to the rear. He looked like a single frame in a Doc Edgerton strobe photo, frozen in midflight as he attempted to reach our yard before a car reached him. I wrapped him in a towel and placed him on the woodpile. Confused and uncertain after years of studying death, I wasn't sure if he was dead yet. Our other cat, the older and wiser Flake, came out on the porch. It was common behavior: Flake sniffing the air, pacing nervously, stopping only when Blitz would appear from across the road. I carried Blitz to the porch and unwrapped him in front of Flake. Flake took a few sniffs of the corpse and continued his search. He had no interest in what I called Blitz. Whatever made Blitz *Blitz* wasn't there anymore. It was safe, I figured, to bury him. Flake continued his sniffing-the-air behavior sporadically through the next few days, apparently looking for his lost friend, then quit and sank into what appeared to be a monthlong depression.

"For all the accomplishments of molecular biology," the biologist Lynn Margulis once said, "we still can't tell a live cat from a dead cat." But we can ask another cat. Who are we to ask about ourselves?

NINE-DAY FRIENDS

Being a hospice volunteer teaches one not to procrastinate. If you have developed a rapport with a patient, and he wants you to come over to talk, or watch a movie, or listen to music, or meet a friend, you go. Right then. The patients at our hospice in western Massachusetts stay with us on average nine days. Then they die, and we don't try to stop them. If you get a lunch offer, it's best not to put it off a week or two.

We get those patients that the medical establishment—doctors, hospitals, nursing homes—have given up on. Therapy designed to keep the patient "healthy" or alive is suspended. A diabetic wouldn't take insulin, for example, nor would cholesterol-lowering drugs be in order. Painkillers, such as morphine, are administered freely. Special requests are honored. One woman asked for French toast at 2 a.m. One of our nurse's aides happily made her a batch. One man, a diabetic, asked for the unthinkable, a giant tumbler of orange juice. Another, with heart and other problems, ordered fried chicken from KFC.

By contrast, a friend tried unsuccessfully to smuggle a beer to his father who was dying in an ICU. The hospital personnel were furious. The man died a few days later. One wonders how much his life was prolonged by refusing him a beer. It is somewhat relaxing to work at a hospice. Our customers are expected to die. Sometimes we fail, though. One of our unit's first patients, when pulled off all therapy and allowed to eat anything he wanted, showed so much improvement that he had to be shipped back to his nursing home, where he promptly died.

The approach of hospice—that death is natural and cannot

be avoided—remains an affront to the medical establishment. Our hospice was investigated because no employee authorized to administer morphine was available for a patient in the middle of the night. Rather than have the patient suffer until morning, an assistant administered the painkiller. The state put our hospice on probation for this violation. After all, imagine what could have happened had the assistant slipped up; the patient might have died.

Even in hospice, though, there are limits to freedom. One cannot smoke, for instance, even outside the building. According to the Centers for Disease Control and Prevention (CDC), the average smoker goes through 500,000 cigarettes in his life. A few more in one's final nine days is probably not going to make a difference. More restrictive are the limits placed by us, the volunteers and staff members. Overtly and unconsciously, we exert our worldview on the people we are supposed to be helping through their final hours. We make sure that even if our patients do not request a moral inventory of their lives, they are provided one.

We don't do this on purpose. In fact, during training we were told that we are not to question a patient's religious beliefs. If those beliefs provide comfort in the final hours, then let them be. One lecturer told us that the slickest deathbed counselors she ever met were clergymen. But the goal of the clergyman is to win souls for his religion, not necessarily to comfort the patient. That is not our mission.

Yet how does one jettison one's beliefs and worldview? I am one of only two male counselors at this western Massachusetts hospice, two oddities among a cadre of dozens of women. Worse, I am a cigar smoker with blue-collar origins, a lover of two-cycle internal combustion motors, and, in general, tragically politically incorrect. One day I got a call from the director.

There was a new patient, Thomas, at the hospice. He was dying, of course, but also "mentally ill." He had taken copious amounts of recreational drugs in his lifetime. He enjoyed heavy-metal music. More troubling, a number of volunteers had complained that Thomas's behavior was often "inappropriate." The committee had met to discuss Thomas's case and decided that I was the most appropriate counselor to deal with Thomas's inappropriate behavior. This was not a good sign. Like telling you that Dick Cheney needs a hunting companion.

Thomas turned out to be a delight. He loved many kinds of rock music. Among other bands, we both admired Cream, the short-lived British electric blues group of the 1960s. An amateur drummer, Thomas could discuss the stick techniques of Ginger Baker (Cream), Keith Moon (The Who), John Bonham (Led Zeppelin), and Mitch Mitchell (Jimi Hendrix Experience). This is not all heavy metal, but to the kindly hospice ladies, I suspect anything left of Mantovani struck them as heavy metal.

Thomas discussed science with considerable knowledge (he had taught it), knew quite a bit about recreational drugs, and was fascinated with Sicily. We were both half Sicilian, but neither of us had visited his ancestral town (Palermo for Thomas, Termini Imerese for me). He said, "Let's go to Sicily!" I said, "Sure," and I immediately went out and bought us a guidebook and several maps to plan our tour. I knew that a trip would probably never take place, but Thomas appeared to *believe* it would happen, and we were told not to mess with a patient's beliefs, so . . . A previous patient believed that he had been taken prisoner and that we were his captors, and he threatened the hospice daily with lawsuits. Then one day he believed he was aboard a cruise ship, and his mood brightened considerably. "These are such nice cabins," he said. "Simple. Not like the last boat I was on." There seemed to be no need to destroy his illusion.

Thomas was several years younger than I was and, despite his terminal cancer, had lustrous long dark (dyed, I assume) hair. He looked a bit like Roy Orbison. He was witty. About his Irish-Sicilian heritage, he said, "I'm genetically cursed with bouts of irrational anger tempered by periods of irreconcilable rage." He told everyone he didn't "belong here." Then where? "Barbados," he said. He asked me if I would write about him.

"What," I said, "like a less maudlin version of *Tuesdays with Morrie*?"

"Perfect," he said. "You be Morrie."

Hospice was not such a good place for Thomas, with its assortment of spiritual books with pictures of serene, conventional people looking skyward and its geriatric ambience. Thomas, who was hard of hearing, and I once tried to watch a rock-and-roll video. Every female hospice worker who walked through the viewing room, whispered, "Mind if I turn this down a bit?" Soon the volume was down to a setting of 3, not optimum for Neil Young.

But what was so "inappropriate" about Thomas that I had to be rushed in to deal with it? I'm pretty sure it was the sex. Thomas was dying, but he was still damn horny. His conversation was filled not so much with innuendo as with direct sexual references. He wondered, for example, if I had slept with either of the very beautiful nurse's aides. I was in my sixties, and they were in their early twenties. (The answer is no.) We talked a great deal about sex, infidelity, love, yearning. Thomas was seemingly proud of a life spent as a "cock master," in his words. He also missed his wife a great deal. She had evidently left him.

Thomas, with his love of drugs, loud music, and sex, was not a proper dying person. Here he was, with a life expectancy of maybe nine days, and he was still expected to toe the line.

Our society has standards, and it never lets go. As I would learn, Thomas couldn't let go either. He told me his beloved but absent wife had red hair and green eyes. I concocted a plan, something concrete I could do for Thomas outside of mere talk. I asked Thomas if he would like an exotic dance by a green-eyed redhead. He couldn't touch her, and she wouldn't be his estranged wife, but she would have the requisite hair and eyes and probably be younger and prettier. She would also dance naked for him. Thomas was clearly very excited. I said I would look into it. As luck would have it, I knew such a woman, and she agreed, without hesitation, to dance for a dying man. She implied that some incidental touching might be allowed. I announced the good news to Thomas and pondered aloud whether I should try to clear the dance through the hospice director or take Thomas to a neutral location, immune to hospice regulations. My dying patient was torn, but in the end he rejected the dance. He said he would "get in trouble." I asked, "With whom?" Thomas moved his eyes to take in the entirety of the hospice. I took this to mean that, in his final days, he did not want to elicit the disapproval of the hospice staff. Even in death, we humans are shamed out of being human.

WILL YOU DIE?

In the 1980s I fell asleep at a meeting, and upon awakening, I discovered that my boss had appointed me the editor of a new magazine, *Longevity.* Our motto, "Some of us may never die," was soon contradicted by calamities suffered by the staff. The publisher, executive editor, art director, and chief writer all died before the age of sixty. The advertising director contracted a mysterious illness[15] and disappeared, as did the magazine itself, to the applause of a grateful nation. I was mercifully fired, for, among other things, suggesting a change in the name to *Morbidity.*

What I discovered during my brief career in the field of life extension is that it is not an undertaking with a serious future. There are two terms that people often confuse. "Life expectancy" refers to how long an individual, species, or class of object can live on average. "Life span" is the maximum time an object or animal can live under ideal circumstances. For some objects, life span cannot be determined.

Take my car. At this writing it is twenty-eight years old. The life expectancy of cars in the United States—calculated by the average age of cars at the time they reach the junkyard—is slightly more than thirteen years.[16] So my car appears to be freakishly aged. But it is juvenile compared to an 1885 Benz, one of the first cars ever made, still in working condition, that resides in the Deutsches Museum in Munich, Germany. With perfect maintenance, it may never die. Cars, and other mechanical objects, thus have a measurable life expectancy but no determined life span. In practical terms, cars are immortal.

The same is probably not true for human beings. Unlike that first Benz, none of the first humans born, say fifty thousand years ago, is still alive. Some of their early tools have survived; but they have not. This is because human beings have a firm life span, somewhere between 115 and 120 years. We can improve life expectancy, the average age we survive to, but as yet no progress has been made with life span. There is a cutoff point. It appears we must die.

When written out, ink on paper, our mortality seems obvious. I've met a few people—like my deceased boss[17] at *Longevity* magazine—who literally have not believed they would actually die. In my dead boss's case, she felt science would find a way to grant her immortality. Her husband, the publisher, once began a sentence, "If I die . . ." But among the rest of the population who admit to their mortality, most are still in denial, as Becker's thesis posits. They live as if death is irrelevant. Is a last-minute reprieve possible for those of us who are still alive?

HOW MANY PEOPLE HAVE ESCAPED DEATH?[18]

Let's take a look at the fate of all the human beings who have ever lived on the planet. Population experts admit they don't know how many people have walked on the earth, given that no demographic data are available for 99 percent of our species' tenure on the planet.[19] Scientists can guesstimate, however, using reasonably speculative birth rates and life expectancies for different eras. One of the problems: when does one start counting? Who were the first humans? One could go back to 700,000 B.C. when the first ancestors of *Homo sapiens* appeared, or even back millions of years to the first hominids. But modern *Homo sapiens* evolved around 50,000 B.C. (probably), and we usually extrapolate from that date. Going back to 700,000 B.C. or before wouldn't make a significant difference, given that there were so few of those individuals, the human/hominid population having not exploded until the past couple of hundred years. The world population didn't pass the billion-person mark until 1800.

The number demographers came up with is 107.3 billion people. Theoretically, that is how many humans have been born on Earth. The number is most likely lowballed. Infant mortality in prehistoric times was estimated at around 50 percent. Scholars admit it might have been much worse and that infanticide might have been epidemic,[20] but let's ignore that for the moment.

To sum up, today, 7 billion people are alive worldwide. So the bottom line is:

107.3	billion total people born on Earth
−7.0	billion still alive
100.3	billion people dead

Death has a big lead. More than 93 percent of all persons ever born have died. About 6.5 percent of us are still holding on. Is the fact that 100 billion out of 107 billion people have died proof that today's 7 billion will also die? We could be special. In the eighteenth century, the Scottish philosopher David Hume attacked assumptions based on apparent stability. He said that the fact the sun has risen every morning since the formation of the solar system is not proof that it will rise tomorrow. But an Englishman, Thomas Bayes, a Presbyterian minister and amateur mathematician, countered Hume's skepticism with a paper, discovered after his death, entitled "An Essay Towards Solving a Problem in the Doctrine of Chances." Bayes's paper said that one could measure one's confidence in a single event happening based on past experience of many such events. We use Bayes's theorem to calculate the probability of any event. We won't go into the technicalities of the theorem here. Suffice it to say that if the sun has come up every morning for 4.5 billion years (the approximate age of the earth), it will probably rise tomorrow. Likewise, if the previous 100-something billion people have died, the present 7 billion are likely to die also.

THE TYRANNY OF LIFE SPAN

Bayes's theorem is just probabilistic mathematics. It doesn't account for technological advances. It would have predicted in 1902, for example, that in all probability, man would never fly. The Wright Brothers shattered that prediction in 1903. There is a better reason to assume our mortality than probability.

The beacon that illuminates our obvious mortality is life span. Automobiles as yet have revealed no life span—no maximum age before giving up the ghost—but human beings have. Our life span is around 120 years. There are claims of people living longer, but none is well documented. A Frenchwoman who died in 1997 supposedly reached 122 years. For those seeking immortality, life span is ominous. It means that no matter how much you floss, how much vitamin C you ingest, or how many colonoscopies you subject yourself to, you will die. Had just one of the previous 100 billion dead humans lived to, say, 5,000 or 1,000 or even 130 years old, there would be hope. It would mean that those yearly mammograms, mole inventories, and regular periodontal examinations might possibly provide immortality.

When I was in the longevity business, it was undergoing a significant change. Beginning in the 1950s, the stochastic model of aging and death began to erode. Stochastic means "exhibiting random or probabilistic behavior." Automobiles age in a stochastic manner, and we can increase their longevity via frequent oil changes, valve jobs, and the like. That approach had often been applied to human beings, but for the reasons just discussed, it doesn't work. Humans age and die in a somewhat predictable, if not well-understood, manner. Such tactics as copious medical tests and diet might delay the stochastic process slightly, but they don't address the basic problem: humans will die, and our life span appears to be locked in. The fact is, *something* will kill us eventually. We tend to ignore this fact and thus are susceptible to hucksterism. In 1970, while researching an article on diabetes, I interviewed a doctor at the American Diabetes Association who told me that in the coming decades, diabetes, at that time the seventh leading cause of death, would rise up the ladder of "dread diseases." The reason? This doctor had concluded that death is inevitable, and as other diseases were treated more efficiently, diabetes was positioned to take their place in the pantheon of deadly diseases. He added that it would be unethical to exploit this phenomenon for fund-raising purposes.

In 2006 *The New York Times* ran a series of atypically overwrought articles on the disease, calling diabetes an "epidemic," "a crisis," with much hand-wringing over its "awful toll." The opening article was scarier than *The Texas Chainsaw Massacre,* beginning with a woman getting two toes sawed off, continuing with a concatenation of horrors that included blindness and deafness, and linking the epidemic to such factors as fewer gym classes in schools, immigration, and cell phones. Hidden deep in the story was another explanation from the CDC: our population is "increasingly comprising older people."[21] That is, as medical science makes inroads against heart disease, stroke, cancer, and other ills, we are living longer and thus become susceptible to other diseases—such as diabetes. The author seemed to feel that it was a paradox that "other scourges like heart disease and cancers are stable or in decline."[22] Exactly where is this paradox? Some deadly diseases decline; others must take their place. Actually, despite apocalyptic publicity from the American Diabetes Association, the disease has moved up only one notch in almost four decades. According to the National Center for Health Statistics, diabetes has moved from the seventh leading cause of death in 1970 to sixth place at this writing.[23]

This is not to deny the importance of diabetes, a disease I have had for several years, or the fact that there are bona fide increases due to factors such as diet among children that has prematurely induced type 2 diabetes in many. But should I and my fellow older diabetics fret ceaselessly about this disease? Dr. Sherwin Nuland reports that there is a more relevant danger.

THE REAL KILLER

Nuland, an MD and professor at the Yale School of Medicine, is the author of *How We Die: Reflections on Life's Final Chapter,* the unsentimental 1994 book that attempted to bring the rough facts about dying out of the closet. "It is not politically correct to admit that some people die of old age," he wrote.[24] When I was a kid, many of my older relatives died "of old age." Today, a specific cause—heart attack, stroke, diabetes, and so on—must be stated. Nuland is saying that "old age" is in fact the underlying culprit; the proximate causes of stroke, cancer, and heart attack are mere epiphenomena.

Nuland wrote that when he attended autopsies, seeking to verify the specific cause of death of a patient, he and the dissector would "tend to ignore the familiar panorama of aging that gradually reveals itself with every added stroke of the knife."[25] While cutting their way to a cancer site, say, or a lethal infection, they would traverse a landscape filled with atherosclerosis and atrophy, which, in their focused search for a specific cause, the examining doctors often failed to notice. Death was just waiting to happen.[26]

At the Yale–New Haven Hospital, Nuland and Dr. G. J. Walker Smith, the director of autopsy service, studied the records of twenty-three patients between the ages of eighty-four and ninety-five. Every patient had advanced disease in the blood vessels of the heart or brain, and almost all had it in both. In other words, no matter what they officially died from, every patient's heart or brain was close to failure. Nuland and Smith found that the patients' bodies were riddled with "incidental" diseases: cancers, aneurysms, kidney problems, urinary tract infections, even gangrene.[27] Those were ills that did not officially kill their host bodies but were waiting in the wings should another disease fail to do the job. "The very old," writes Nuland, "do not succumb to disease—they implode their way into eternity."[28]

EMPTY CHAIRS IN THE SENIOR CENTER

When I turned sixty, I signed up for a discussion group that met weekly at my town's senior citizen center. We discussed such things as evolution, the environment, conspiracy theories, futurism, faith, and healing—generally topics both scientific and spiritual. Our members included a doctor, some psychotherapists, teachers, a human resources specialist, a nurse, two actresses, a CIA analyst, two musicians, and a few scientists, one a member of the National Academy of Sciences. The discussion group had a formal name, but some of us referred to it simply as the Geezer Group.

There are twenty chairs available in the meeting room at the senior center, but one can always get into the group because the chairs are vacated at a furious pace. It might be announced that "Beth Ann won't be here this week; she's not feeling well," to be followed a few weeks later by "Beth Ann's memorial service will be held . . ."

We geezers are cautious. When a new activity—say, a workshop for psychic phenomena or exercise is suggested—members always cite the dangers. One new member said that she felt a pall of fear envelop the room. At my suggestion, we devoted a session to our fear of death. It is less rational for us, one would think, because we have so much less to lose. If we mess up and get ourselves killed, we are risking only a few years of our potential lives, as opposed to twenty-year-olds or younger, who put sixty-plus years on the line every time they take a major risk. We are "those about to die." We should be fearless, enjoying our lives to the fullest.

At this particular session, members said they had no fear—because they weren't going to die. Okay, no one said these words exactly, but the men bragged about how often they had colonoscopies, prostate exams, full-body scans, et cetera. I appeared to be the only man who did not believe that a fiberoptic tube stuck up my behind guaranteed immortality. The women were less into technology and more into vitamin supplements, including heroic doses of fish oil. They were *healthy.* They seemed not to have noticed the rapidly emptying chairs. Then there were those who took another angle on their fears. The CIA analyst, his girlfriend, and another Christian lady said death didn't matter because they were going to heaven. Fear triumphed in the end when one woman told of her neighbor who had died from a colonoscopy. The men looked nervous. The women kept gobbling down fish oil capsules.

WHY DO WE DIE?

During my brief foray into the science of longevity, I learned that serious researchers had begun to jettison the old stochastic theory of aging. Comparing people to automobiles or other machinery was proving less than fruitful. As mentioned earlier, the "business" of longevity in the past few decades has seen an effort to target life span rather than life expectancy. Leading scientists in the field shared Nuland's belief that mortality is an inevitable companion of aging and chipping away one disease at a time is futile, like rearranging deck chairs on the *Titanic.*

It began to appear that living organisms don't wear out willy-nilly but are somehow programmed to die. To live is to die; life is an inexorable journey toward death. Longevity research took a sharp turn, and instead of examining medical and dietary techniques, researchers began to ask what causes death itself and to seek out ways to extend life span instead of just padding out life expectancy.

The most extreme theory was the "death hormone," a theoretical—and yet undiscovered—substance released by the body that actually kills us. The prime exponent of this hypothesis was W. Donner Denckla, an endocrinologist, who hoped to isolate the hormone and then design an antidote to defeat it. He was unsuccessful.

There were other theories explaining how death comes about, how it is programmed into our genes or at least is an inevitable consequence of the way the human animal (and others) is wired. Though research has not supplied an explanation of death, it has changed our approach. We have accepted for centuries that death is the default. To die is what happens and would naturally occur to a living thing. Work by Denckla and others reversed the approach. Perhaps death is the unexplained phenomenon; perhaps immortality is the default. Organisms would live forever were it not for some secret mechanism that kills them.

The pivotal finding in longevity research is what's now known as the Hayflick limit. Until 1961 biology had virtually nothing to say about aging.[29] The standard view was that cells were immortal. In a series of sloppy experiments conducted at the Rockfeller Institute in the first half of the twentieth century, chick cells placed in a culture dish continued to divide unabated when doused almost daily with nutrients. No one is sure how the experimenter, the Nobel Prize winner Alexis Carrel, obtained this erroneous result; one possibility is that when nutrients from freshly killed chicken embryos were added to the cultures, Carrel and his colleagues were inadvertently supplying new cells each time.[30]

Leonard Hayflick changed all that. In 1961 he published a paper that revealed that cells are mortal. In fact, they are almost precisely mortal, the now-famous Hayflick limit being fifty. In twenty-five separate strains of cells, Hayflick demonstrated that a cell will divide and replicate about fifty times, at which point it hits a wall—and stops.[31] The Hayflick limit, which holds sway over the 100 trillion cells[32] in our body, has a lot to do with why we are not like a Benz automobile.

But why should a cell limit itself? One of the theories put forward, by the gerontologist Alex Comfort[33] and others, was the "bad photocopy" hypothesis. One photocopies a document, then it is recopied, and after fifty times it's a mess—the same thing that happens in games of "telephone" or "Chinese whispers," in which a statement is passed from one person to another, with either hilarious or disastrous results.

A modern, and more sophisticated, theory has to do with telomeres, the very ends of the chromosomes in our cells. In bacteria, the chromosomes are circular. A circle has no ends and bacteria can reproduce ad nauseum. Our chromosomes are linear, sticklike. In 1966 the biologist Alexey M. Olovnikov began work on the "end-replication problem": every time a cell divides, a bit more of the end of the chromosome fails to get copied. The problem struck Olovnikov as he waited for the subway after hearing about Hayflick's work. I quote Olovnikov's epiphany from the definitive book on the topic, Stephen S. Hall's *Merchants of Immortality: Chasing the Dream of Human Life Extension:*

> "I heard the deep roar of an approaching train coming out from the tunnel into the station itself," he recalled. "I imagined the DNA polymerase to be the train moving

> along the tunnel that I imagined to be the DNA molecule." He realized that if the track represented DNA, and the train was the enzyme that copied it, the locomotive would be the front of the enzyme that pulled the copying machinery but was itself incapable of copying the DNA directly beneath it. After fifty such copying operations, Olovnikov figured, the chromosomes would become unstable.[34]

To which I say, I'm glad Olovnikov wasn't waiting for a taxi or JetBlue. God knows what kind of metaphor we'd have to deal with. In any case, the telomere problem is the new "Xerox problem." In all seriousness, this is reasonable work, and something good may come out of it, though immortality, pushed by the theory's proponents on talk shows (who never look very healthy, by the way), may not be as prompt in coming as the next subway train. Still, given Hall's reputation for responsible reporting, it worried me that immortality might possibly be around the corner, so I wrote to him asking his prognosis. He wrote me back a reassuring note: "Death," he said, "continues to be a growth industry."

In any event, whether we call this new theory "the Xerox problem" or "the telomere problem," we're still back talking about a stochastic process. If the chromosomes had been copied correctly, King Tut might still be alive.

EVOLUTION

What does evolutionary biology say about death? The alleged beauty of natural selection is that species select for traits that increase their chances of survival and select against those that decrease their odds of survival. What could be less advantageous than an organism that dies? What could be more advantageous than a species that evolved a better technique of copying telomeres, or whatever it is that would eliminate death?

Most biologists harrumphed when I asked them, "Why wouldn't immortality be an obvious outgrowth of natural selection?" One who took on the question, reluctantly, is Theodore Sargent, a retired biologist from the University of Massachusetts at Amherst. He said that immortality would pose some problems: (1) An immortal population would be predominantly old and at a disadvantage whenever youth was favored. (2) Longevity might give some competitive advantage in the short term, but the immortal population might be unable to take advantage of sexual reproduction, which provides diversity, to survive climatic shifts and other disasters in the long run. (3) The old might squeeze out the young—"if only because of physical limits to the numbers of organisms that can occupy a given space." Sargent went on to say that this could lead to what demographers refer to as a "senile, declining age-distribution" profile, that is, weighted toward the "old"—like the United States today.[35]

These are all seemingly logical, evolutionary arguments against immortality. In the short run, not dying would be advantageous. In the long run, immortality theoretically causes problems. As Sargent concluded, "I suppose I'm saying that immortality 'experiments' over the long haul may have led to extinctions—and therefore no extant examples." In fairness to Sargent, I was putting him in an awkward position, asking him what arguments evolutionary biology might put up to justify death as an advantageous trait for a species—how not surviving would benefit survival. Other biologists would not tackle the problem because it is thorny at best, embarrassingly fatal to our modern worldview at worst.

The problem is that each of the arguments against the

advantage of "not dying" is teleological, which is to say goal-oriented. The Greek root of "teleology," *telos,* means "end." Something that is teleological is thus directed toward a definite end or has an ultimate purpose. It requires natural selection to be godlike, to be able to predict the future. Natural selection is supposed to be a simple matter of random mutations. Most of those mutations are harmful to the organism, but in rare cases an advantageous mutation occurs, say, a curved beak that allows a bird to dip into a flower for nectar or a flatter tail that provides a beaver with speed and mobility. Those mutations are selected for, goes Darwin's respected theory, and become commonplace in a species, helping it to survive.

A curved beak might one day become a disability, but natural selection has no way of knowing that. Natural selection is like a desperate corporate CEO, worrying about third-quarter profits only and not thinking about ruination down the road. Some species outgrow their food supply and perish, but that does not keep them from evolving toward better reproduction and food-gathering abilities despite the eventual dismal outcome.

Sargent hints that there is missing evidence. Have there been "immortal" species that died out for the express reason that they were immortal? Perhaps, but no such fossils have been found. (By "immortal species," I mean a life-form that does not die "from the inside out," from a deteriorating body; it would still be subject to death from trauma—such as falling off a cliff, drowning, or being shot.)

We like to make sense of things. We like there to be a nice, smooth narrative to give meaning to our lives and deaths, but there may not be one despite the efforts of neo-Darwinists. Jean-Paul Sartre wrote that if God is dead, it leaves a God-shaped hole in the universe. People fill that hole with whatever's handy, and in the case of many modern biologists, they fill it with natural selection, a godlike causal process that we hope will explain all of life's mysteries, including death. We may be asking too much of natural selection, which has plenty to do making beaks bend in the proper direction.

SENESCENCE

A very different kind of evolutionary biologist has different ideas on death. I last saw Robert Trivers at a cocktail party in his honor at the home of an Amherst College professor. Trivers had just received an honorary degree from the college. The scientist is always a good interview, having no internal censor.

A couple of decades ago, one of my writers at *Omni* magazine, Bill Lawren, interviewed Trivers for a formal Q and A. At the end of the process, Trivers turned to Lawren and said, "The big question here, Bill, is: how much pussy is this interview going to get us?" Spoken like a true Darwinist. Between that time and the cocktail party, Trivers had gotten married, was dealing with melanoma, had mellowed a bit, but was no less blunt. "Teresi!" he yelled at me from across the room. "Have you heard about the Atkins diet?" Trivers was pointing out that I needed to shed some weight, but in fact he also really wanted to discuss the Atkins diet, which he was following—inexplicably, in my mind, because I thought he was too frail to begin with. Trivers seems obsessed with personal attractiveness. It's possibly an occupational hazard. He studies, among other things, mate selection: what makes a female attractive to a male and vice versa. *Omni* magazine was owned by Bob Guccione, also the publisher of *Penthouse* and a master photographer of naked women, and I had worked for Bob for eleven years. Trivers had a theory that breast size was irrelevant to the attactiveness of human females, that it was symmetry that attracted males. He asked if I could introduce him to Guccione so that he could measure the centerfolds to determine breast symmetry.[36]

With the death of his friend and collaborator W. D. Hamilton, Trivers remains the most accomplished and most important of the modern neo-Darwinists. Much of what you read about modern evolutionary studies sprang from work by Trivers and Hamilton or has been blatantly ripped off. In the 1970s Trivers wrote five papers in which he applied genetics to behavioral biology, including bird warning calls, cuckoldry, revenge, and sibling rivalry. His Harvard colleague E. O. Wilson, without giving full credit, fancifully popularized Trivers's work when writing the book *Sociobiology,* which made Wilson a world-famous figure. Even so, Trivers spoke highly of Wilson to me, only complaining that Wilson had jumped the gun in applying the research to humans. Recently, though, he has shown more pique, telling *The Boston Globe* that Wilson made himself "the father of the discipline, when he's really the father of the name of the discipline."[37]

Trivers is a strict neo-Darwinist, believing in natural selection as a kind of perfect engine for evolution, fastidious in its ability to select for advantageous traits in life-forms and to select out disadvantageous ones. He believes there is no such thing as "waste" or random DNA. Every gene has a purpose; nothing is there by accident.

So I asked him, why do we die? How could natural selection mess up so badly? As mentioned, what could be a better trait for survival than not dying? I thought it was a simple question. He sucked at his drink, preferring, I suspect, to discuss mating strategies or to get back to the Atkins diet. Finally, he said, "Can't help you there." He knew of no legitimate work that explains the popularity of death in four of the five kingdoms of life.[38] It was hard to imagine that I had stumped Bob Trivers.

After a few more sips, he said, "Maybe you're asking the wrong question." Trivers's solution was not that natural selection had made a clumsy mistake in selecting for death but that it had chosen aging, with death as an inextricable consequence. He sent me back to the master, W. D. Hamilton, and his theories on senescence.

Hamilton, an Oxford biologist, was an important person for many reasons. For one, he rode a motorcycle and refused to wear a helmet. His judgment proved sound since he died from malaria in 2004, a disease unaffected by helmet use. He was not a great popularizer or, reportedly, even an effective speaker, sometimes using his microphone as a pointer during slide presentations so that no one could hear him. But lesser scientists (e.g., Richard Dawkins and E. O. Wilson) made fortunes from popularizing his work.[39]

His greatest contribution was changing the way scientists looked at natural selection and evolution. Why, for instance, do many animals exhibit altruism? Why do they put themselves in harm's way, sacrifice themselves for the sake of others? Hamilton's answer was simple but revolutionary: the living animal is not important; it is not the instrument of evolution. It is the gene, or a copy, that goes on to the next generation. That is why we see altruism, goes this theory, for instance, in bees. Why do the workers care for the queen's offspring rather than breeding themselves? Why do auntie bees commit suicide in defense of the hive? Hamilton said that bees are so closely interrelated that they are working or fighting for the propagation of their own gene pool. That is supposedly why parents will die to save a child. The theory is called "inclusive fitness." Decades earlier the geneticist J. B. S. Haldane had touched on it by saying, in a pub, "Would I lay down my life to save my brother? No, but I would to save two brothers or eight cousins." The joke is that one shares only 50 percent of one's genes with a brother and 12.5 percent of genes with a cousin. You need to gather enough relatives to break even.

Hamilton's gene's-eye view of evolution leads to a rationale for senescence. Hamilton called it "Live now, pay later." In his paper "The Moulding of Senescence by Natural Selection," he quotes Temple, a character in James Joyce's *A Portrait of the Artist as a Young Man,* who states, "Reproduction is the beginning of death." Hamilton's paper is mostly algebra, but Trivers explained that he was leading up to a simple idea: genes have their best chance of being passed on if the individual is wildly sexual, reproducing at a furious rate while young and not pacing itself. This leads to senescence. The body deteriorates later with age. Reproductive success and aging are inextricably linked, but in a reciprocal manner. A wild sexual life in one's youth does not explain death per se, says Trivers and Hamilton, but it results in senescence, of which death is a faithful companion. (We are speaking here in a specieswide sense, not individual by individual. A virgin will not achieve immortality.)

These concepts, developed by Hamilton and Trivers and popularized by Wilson, Dawkins, Daniel Dennett, and many others, are accepted under the rubric of neo-Darwinism. Most of the evolution you read about in the press or hear about on television is of the neo-Darwinist bent, which holds that natural selection leaves nothing to chance, that there is meaning in every beak and feather in nature, in every social behavior from altruism to mating to love. It is all shaped by natural selection. There are algebra and reasoning to explain why bees act the way they do and why animals get old, but the discipline is often lacking in evidence in the field, the laboratory, and the fossil record. The theories of Hamilton, Trivers, et al. are appealing and comforting, reimposing Newtonian determinism on a universe in which such determinism was discarded more than a century ago in favor of quantum randomness and uncertainty. Sargent points out that Hamilton's ideas about death verge on circular reasoning and at the very least remain unsatisfying.[40]

A very different kind of evolutionist was fond of mocking the neo crowd and had her own theory about why our time on earth is limited.

SEX KILLS US

On a dark night, I almost killed one of the world's most important scientists. I was pulling out of the high school parking lot when my wife shouted, "Look out!" I braked as a dark figure on a bicycle zipped by in front of my grille. It was Lynn Margulis, a University of Massachusetts biologist and member of the National Academy of Sciences. Margulis was a familiar figure in our town, an inspiration to cyclists and geoscience majors alike, but one wishes she didn't wear a black jacket while biking at night. Her tiny Japanese headlight was not the beacon she professed it to be. She went almost everywhere on the bike, even in winter, gussied up at night for formal campus dinners.

She turned up at odd times in my life and was considerate of my developmental prosopagnosia (I don't recognize human faces) by repeating her name if she was not on her bike (I do recognize bicycles). This happened once at 11:30 p.m. at the Super Stop & Shop. We both liked to shop around midnight to avoid the crowds, and bumped into each other in the produce aisles. She said, "I don't understand people who become vegetarians because they believe in the 'sanctity of life.' Sponge off your kitchen counter, and you've just killed fifty thousand fungi." I dumped my carrots and Belgian endive back on the shelves, pushed my cart to the meat department, and bought the biggest rib roast I could find. Thanks to Margulis, I now eat plenty of meat, but I don't sponge off the counter.

A Distinguished University Professor in the Department of Geosciences at UMass, Margulis insisted that the study of evolution extends beyond academic biology, requiring knowledge in all kinds of geology, such as paleontology, oceanography, atmospheric chemistry, and all the environmental sciences. She took her message wherever she could. One Sunday I found her giving the sermon from the pulpit of the local Unitarian Universalist Meeting Hall. She began by stating she hadn't been in a church in forty years, then went on to say that if a bacterium were allowed to reproduce, unimpeded, within one week there would exist a clump of bacteria whose total mass would equal that of Earth. We never wake up in the morning and see Earth overwhelmed with bacteria because natural selection keeps that from happening. Charles Darwin, who was more conversant with large sexually reproducing birds and mammals, did a similar math exercise with elephants in the nineteenth century. He calculated that elephants would cover Earth with the progeny of a single pair within a century or two if unimpeded. His arithmetic has been criticized, but his lesson was the same as Margulis's: the populations of all species, including, of course, humans, have a tendency to outgrow their environments. This maximum size any population can reach, measured in the largest numbers of offspring per generation possible for that species, is called "biotic potential." This maximum, e.g., thirty-two children per couple or 99 × 12 dachshunds per generation (33 puppies per litter × 3 litters a year for 12 years) is almost never reached because of natural selection. Many offspring are born, hatched, budded, or sprouted from seed but few survive.

It was a wondrous thing to hear Margulis delivering the discouraging mathematics of reproduction and survival from a pulpit. To go from sperm or egg to fertilization to implantation to fetus to birth through infancy and into old age requires one to beat staggering odds. There in the Unitarian meetinghouse, I saw the telltale sign of men who had beaten those odds. Most of the male parishioners had, like myself, crossed the sixty-year mark, some long before. Standing in the last pew, I witnessed the evidence. No matter how big or small their stomachs, few of the men could fill the seats of their trousers. The butts of old men tend to disappear, and one wonders if human biologists can explain this phenomenon. What's more mysterious is the fact that every Sunday, churches across the land are filled with sixty-year-old-plus people, who despite their unbelievable fortune of having attained life and then a highly improbable age, are praying for even more years, to be "saved" and given another life beyond this one. No one in those pews is going to survive. One of the problems is sex.

Many of us, depending on age, were taught that there are two kingdoms of life: animals and plants. Today, it is accepted that there are five kingdoms, and Lynn Margulis, drawing on work by natural historians dating back to the seventeenth century, was instrumental in the recognition of this biodiversity. The kingdoms are bacteria, protoctists, fungi, plants, and animals.[41] But a case can be made—and Margulis made it—that there are really only two kingdoms: bacteria and everybody else. Bacterial cells (they are called "prokaryotes") never have nuclei, whereas all other life-forms have at least one nucleus per cell (they are "eukaryotes"). Bacterial cells contain their genes in skinny tangled threads of DNA whereas everybody else has genes padded with lots of protein on lengths of protein-studded DNA called chromosomes that are packed inside their nuclei. The chromosomes stain red in eukaryotes whereas they can't even be seen in live prokaryotes, stained or not. Margulis's 1965

PhD thesis was the starting point for her argument that small bodies inside cells such as mitochondria and plastids evolved from bacteria that were long ago incorporated into cells. She helped explain the evolutionary leap from prokaryotes to eukaryotes. The late Ernst Mayr, the greatest biologist of his time, said, "The evolution of the eukaryotic cells was the single most important event in the history of the organic world . . . and Margulis's contribution to our understanding the symbiotic factors was of enormous importance."

One cannot view the leap from bacteria to nonbacteria as "progress." Many textbooks, as well as the media, retain an ancient Greco-Christian "ladder-of-life" view, a pyramid or tree with man at the top and every other being down below, bacteria at the bottom. That view is a religious legacy, not science. According to Margulis, "All extant species are equally evolved. All living beings, from bacterial speck to congressional committee member, evolved from [an] ancient common ancestor. . . . The fact of survival itself proves 'superiority,' as all are descended from the same metabolizing Ur-form."[42] This is an important scientific point for any discussion of the death of people. As we shall soon see, medical science continues to put humans into a special category, with ad hoc criteria for our deaths that separate us from all other species.

The evolutionary phenomenon that concerns us here is the move from asexual (one parent only) to the kind of reproduction we mammals have that requires two parents, one father and one mother. (Many, many other kinds of sexual fusions exist, including some that are not directly related to reproduction at all. Indeed, some kinds of life have four, eight, or even more than a hundred different genders. They are just not animals.)

Animals need sexual fusion of their parental cells to precede reproduction (making babies), and this, said Margulis, makes "programmed death" a necessity also. Outside of the fact that we animals get to have sex, sexual reproduction has no advantages. Margulis said that sex is simply *required* to make all animals. Some scientists like to insist that sex combines the DNA of two parents and thus creates diversity in their offspring. This is not true. Many necessarily sexual organisms—take cheetahs—lack variation, while cloned tomato cells, all from the same original parent cell, display all kinds of genetic variation. Margulis said that biologists are full of rationalizations for processes they don't understand and fill in the gaps in their understanding with unstated prejudices. The sexual act in animals, as Margulis saw it, "is to be attracted to each other, open their bodies, open their membranes, in this case a sperm and egg. Fuse their contents instead of one eating and digesting the other." But a problem thus arises. You have two parents combining their DNA. "You have two of everything," Margulis said. "You have to get rid of the doubleness. You understand? You have to have a programmed way of getting the oneness back because otherwise, next time you double and double, you'd have a monster." Fertilization is the merging of two to make one, so there has to be a reduction, a return to "oneness," as Margulis put it.[43]

Hence cell death. The cell is the basic unit of life, and the death of the cell is the gold standard for the death of animals, particularly mammals, of which humans are just one of about eight thousand species. Cell death was introduced on Earth via the second kingdom, the protoctists. The introduction of sexual reproduction mandated cell death, the destruction of the extra DNA (genes enclosed in nuclei) and all the other cell material (nuclei, mitochondria, and so on). We humans are animals, and as such we are relative newcomers to the earth. So is death. Perhaps, said Margulis, it is only natural that we resist and deny death: "Only accidental, externally caused death existed at the origin of life. So it was for a long time thereafter."

I interviewed Margulis several times in her Amherst, Massachusetts home. She always made me yerba maté, strong tea favored by gauchos, and there were always people coming and going who lived in her house: students, visiting scholars, friends, friends of friends, and anybody interesting who needed a place to stay.

Her house sits next door to the Emily Dickinson Homestead. Like Margulis, Emily was fascinated with sex and death, which were inextricably conjoined in her poetry. Margulis was fascinated with the poet, having gotten caught up in the debate over whether Dickinson died a virgin or secretly (during a certain limited period of her life) fucked like a bunny. Having examined her poetry and read recent decodings of same by scholars, she took the latter position.

Margulis was concerned with what is alive and what is not. There is something very different about a live being compared to a nonliving object. She pointed out an obvious fact: we put great importance on DNA, "the master molecule of life," but Margulis said DNA per se is not alive. It is a long, skinny chemical.

"What is life?" is perhaps an unanswerable question. But death in living things may be no less mysterious. Margulis called death "the great perplexer." Many of our ancient and prehistoric ancestors, she said, did not see the universe as mechanical, with the sun, moon, and planets moving according to Newtonian (or the Eastern equivalent) principles. Celestial bodies were often seen as animated by spirits within. Those spirits were apparently immortal. The wind continued to whistle, the changing phases of the moon continued unabated, the sun revolved in the sky. But humans fell over and died. Is it any wonder that people assumed that some kind of spirit or soul that animated their dead friends' bodies had departed? Or, as the nurse at Baystate Hospital in Springfield, Massachusetts, put it, "Whatever made her *her* is gone."

LONG, DREAMLESS SLEEP

I became interested in death one evening in 1996 because of Carl Sagan,[44] the planetary astronomer and premier science popularizer of his time. Every science writer and editor during the 1970s and '80s owed a debt to Carl Sagan. He was the rising tide that lifted all our boats. Mine was a particularly leaky raft. In the mid-1970s, I was the editor of *Science Digest,* a struggling little magazine, and Sagan wrote the cover story for our July 1976 issue about the Viking Lander, which touched down on the Martian surface that month. Our single-copy sales literally doubled.[45] Sagan also wrote for me when we were launching *Omni* in the late 1970s and early 1980s, giving the magazine much-needed credibility. He appeared twenty-six times on *The Tonight Show with Johnny Carson* and became an international superstar with his own TV show, PBS's *Cosmos.*

Sagan was more striking, more handsome, in person than he appeared on television. That's why his appearance on a December 1996 evening on *Charlie Rose* was so upsetting. He told Rose that he had been "saved by science," but he looked anything but saved. He was only sixty-two, yet his appearance was of someone in his seventies or eighties, as he had been ravaged by myelodysplasia, a bone marrow disease. He had been treated at the Fred Hutchinson Cancer Research Center in Seattle with a bone marrow transplant. It had not succeeded, and he had undergone a second and then, according to his widow, Ann Druyan, a third. My wife, who had cancer and had spent time with terminal patients in cancer support groups, said, "He's going to die." It wasn't just his appearance but his stereotypical affect of denial.

When the show ended, we discovered that Sagan had already died. We had been watching a rerun from a few months past. Early in his career, Sagan was known for his open-mindedness, championing the search for extraterrestrial intelligence (SETI). Later he became a skeptic, one of the founding members of the Committee for the Scientific Investigation of Claims of the Paranormal (CSICOP). He professed to be a devout atheist. Sagan had a particular crusade against alternative medicine, especially unorthodox cancer therapies. His second and (possibly) third transplants were what one doctor called "celebrity transplants." An ordinary person wouldn't get one because of the expense and the unknown chances of success. When the first bone marrow transplant failed in a patient older than sixty with myelodysplasia, there was no evidence in that era that a second or third could succeed. The Fred Hutchinson Center could provide me no statistics to the contrary.

Had Sagan's skepticism and logic failed in the face of death? I don't know. Perhaps, like Roger Staubach,[46] he was just throwing a "Hail Mary" (though I think he would have hated that term) pass because he felt he had no other options. It is the same path taken by patients who explore alternative medicine when the only other option is acquiescence to death. I prefer to think of Sagan as fighting the odds rather than giving up on his philosophy.

Sagan did show a softening of his hard-edged scientific approach when he told Charlie Rose that he thought death was "a long, dreamless sleep," alluding to a line from Shakespeare. We have learned some things about sleep since Shakespeare, and it has little in common with death. The heart continues to beat, you continue to breathe, many brain activities continue. If only Sagan were right, it would be a lovely way to go.

Most important, you would not be dead. You would still exist. One of the frightening things about death is that the self is annihilated. We are gone. Like all of us, Sagan appeared to be looking for a way out by telling himself that he would just be sleeping.

Clearly it was difficult for Sagan, it is difficult for me, and probably difficult for most of you to accept the fact that we will be annihilated when we die. It is a tough idea to struggle with, one that the existentialists say most of us never come to face. I brought up Sagan's death only to point out that in the face of death, we turn our backs on hard-won logical principles. And why not?

WHERE WE ARE GOING

As mentioned, I began this project as a magazine article.[47] I set out to explore how the determination of death was now aided by space-age technology. My assumption was that today we have diagnostic equipment that can signal a bright line between life and death. My assumption was that the inexact art of death determination that had confounded doctors and others for millennia was a thing of the past, that modern gizmos and gauges had replaced crude instruments, old wives' tales, superstitions, and religious beliefs about when the soul departed. I never finished the article because after just a few months of research, it became apparent that (1) few doctors are using high-tech equipment, and (2) it doesn't work.

Not that there hasn't been progress. Medical scientists have shown in the past few decades that life is far more tenacious than we knew. The moment of death, if there is such a thing, may come long after it's formally declared. The disconnect between when attending doctors declare a human dead and when more exacting medical scientists think death actually occurs can be roughly depicted by the unscientific, not-to-scale line graphs below. "A" is the point at which doctors or others tend to declare a patient dead; "B" is when more careful overseers thought or think death actually occurs.

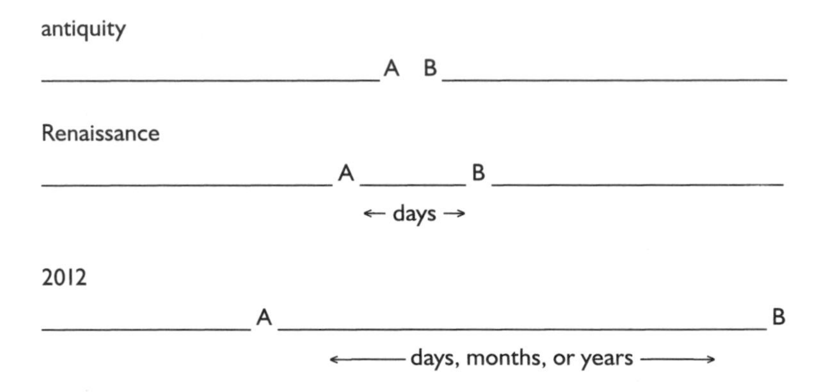

Don't take the above too literally. It merely shows, in a general way, that the gap between the time at which it's legally or socially acceptable to declare people dead and the time at which more enlightened experts in the era actually think death occurs has widened through history. Much of this book will be about the line segment AB, a kind of *undead zone.* This gap is imprecise in modern times because of the range of remarkable case histories. Note that B has moved far to the right, but A has also moved to the left as we rush to declare death. People declared dead come back to life with some frequency. Or an organ donor begins to breathe on his own as his liver is being removed. We can discount some cases as the mistakes of inexperienced medical personnel. Other cases are more problematic. Pregnant women who are legally dead, for example, have been sustained on life support for a hundred days or more to save their unborn babies. This is now being done with regularity. These women are dead yet act alive: their hearts beat; their lungs breathe; they urinate; they get morning sickness; they gestate and deliver healthy babies. In one remarkable case, a brain-dead boy was kept in a similar condition for twenty years. Death is not what it seems.

I thought I had chosen a simple topic. Who is alive? Who is dead? I assumed that science held the answers. Instead, I found that bad science was being used to obfuscate the distinction between the living and the dead, while good science, which was finding that life is far more persistent in humans than we had imagined, was being ignored. Patients suffering strokes can now be saved but can just as easily be judged to be dead by our criteria. The same is true of locked-in syndrome, persistent vegetative state, coma, and other states. Whether you are judged to be living or dead depends on which doctor gets to you first.

As a science writer, I had hoped to write about the science of death determination. But the denial of death affects scientists, too, so we must discuss the strange secular religion they've adopted in their work.

WARNINGS AND PROMISES

I had some difficulties working with this material over the past decade. I left it and then returned several times. I would awaken early some mornings afraid for my death. What was bothering me? The process of death? Or death itself?

One of the benefits of thinking about death for several years is that I stopped worrying about many, many things. A mundane example: My roofer asked whether I wanted to buy thirty- or forty-year shingles. He recommended springing for the forty-year. I said, "Why? So that my roof will last twenty-five years after my death instead of just fifteen?" The roofer thought that was a strange way of looking at things. It's strange only if you don't plan to die. Clearly my roofer had no such plans.

I never quite cottoned to existentialism when I was young, but it makes sense now that I'm in my midsixties. Meursault, the protagonist of Albert Camus's novel *The Stranger,* doesn't fully appreciate the beauty of his life until he is condemned to death. Then he decides that one time is as good as another to die. That was not how I felt at age twenty-one as I made my way through the stations of the army draft physical.[48] It did not seem to be a good time to die. Vietnam did not seem a good place to do it. After surviving the horrors of middle school, the tedium of high school, and the tuition of college, I was not ready to pack it all in without getting some return on my investment. Today existentialism seems appropriate. Does it really matter whether I live to sixty-eight, seventy, or the U.S. average of seventy-six years old? I suspect that most of my age cohorts would disagree.

A VERY QUICK PRIMER ABOUT THE BRAIN

In the following chapter, we will discover how the brain, of little importance to some ancient peoples, became more important in man's determination of who is alive and who is dead as we get closer and closer to the modern age. In chapter 3, we will see that it became all-important, the rest of the human body relegated to an afterthought. Though the term "brain death" is constantly mouthed by declarers of death, the entire human brain is of little relevance. It is the brain stem that counts.

The spinal cord meets the brain at the stalklike brain stem, a grand thoroughfare for sensory and motor signals. The brain stem's business includes such basics as breathing, sleeping, and waking. On a primitive level, the brain stem contains the on/

off switch of consciousness. Attached to the back of the brain stem are the cerebellum (the "little brain"), which generally controls motion, and above are the thalamus, hypothalamus, and pituitary. There are other structures, but sitting atop all this is the largest, two-thirds of our brain's mass, the cerebrum, its two hemispheres draped imperiously around all the other parts. Covering the cerebrum is the wrinkled crust of the cortex, sometimes called the neocortex to emphasize its evolutionary newness. It's a folded sheet of tissue about three millimeters thick that covers about one and a half square feet if unfurled. The cortex made its first significant appearance in mammals. In humans, it is vast, containing at least 70 percent of the central nervous system's neurons.

Snakes, for instance, do not have a cortex and thus do not plan for the future, worry much, or solve differential equations. Dogs, cats, and mice all have one, and they can learn from experience, master mazes, and anticipate reward or punishment. Whatever you may think of humans, we are well endowed cortexwise. As the two Nobel Prize–winning neuroscientists David Hubel and Torsten Wiesel put it, "A mouse without a cortex appears fairly normal, at least to casual inspection; a man without a cortex is almost a vegetable, speechless, sightless, senseless." As we shall see, because our brains are considered "better" than other animals', its loss has come to signal our death no matter how healthy everything below might be. Unfortunately, as we shall also see, it is not the "special" part that is tested in matters of death determination, but rather the more primitive brain stem.

Another problem to be aware of is that neurologists and neurosurgeons, who have taken the lead in deciding who's dead and who's alive, can be talented healers but are not always the best scientists. The move toward the brain as the organizing organ in the body came to a finale in the 1960s, and the debate is still largely framed in the terms of that decade. There has been a lot of neuroscientific water over the dam in the ensuing years, however. In 1973, Candace Pert and Miles Herkenham discovered opiate receptors in nervous tissue. An opiate receptor is a site to which natural opiates, which turned out to be endorphins, attach. Pert's discovery kicked off a brain revolution. Dozens of other receptors and neurochemicals have since been discovered, and the brain suddenly became more fluid. Neuroanatomy has also broken down. The distinct boundaries between the brain parts have gotten fuzzier. Yet now neurologists and others in the death determination business still see the brain in Tinkertoy Newtonian terms, a knee-bone-connected-to-the-shinbone-connected-to-the-ankle-bone mentality.

Pert says that since 1995 a new paradigm has taken hold in neuroscience. Pert, the chief scientific officer of Rapid Pharmaceuticals, which is developing drugs for AIDS, pain, Alzheimer's, and other conditions, says the brain should be viewed as a field—much as Michael Faraday revolutionized physics by visualizing electricity as a field, rather than as a jumble of distinct components. "The regions of the brain are well connected," Pert says. "Just because one area is dead is meaningless. Just because the brain stem is dead doesn't mean the cortex is dead."[49] Trying to discuss such concepts with members of the death determination establishment is difficult. "What do receptors have to do with it?" one neurologist harrumphed at me. If our discussions here about the brain seem at times outdated, there's a reason for that.

A final caveat before we continue. I am writing as a science journalist, not as a philosopher. I am not trying to impose my personal beliefs upon the reader. You may want to believe that when it comes to declaring you or your friends and relatives dead, you are in good hands with the medical community. It's fine with me if you adhere to that belief even though the facts I have uncovered would seem to undermine such a conclusion. My job as a journalist is to reveal data. You can do with that information what you wish. The goal of the journalists who reported on the explosion of the *Hindenberg* was not to undermine the future profits of the zeppelin industry; it was nevertheless their job to report that the docking had gone poorly.

I adore Western medicine. I trust my doctor with my life. I'm just not sure I trust her with my death. Keep in mind that when it comes to your body and those of your family and who's dead and who's alive, who's conscious and who's not, your own judgment may be better than anyone else's.

TWO

A History of Death

Telling a dead human from a live human has never been easy.

I LIVE on the border of a wildlife preserve. It is not as idyllic as it sounds. On nights when I am afflicted by insomnia, I hear hideous screeches, cries, and shrieks emanating from the woods. There are the fierce growls of predators and the suffering moans of the wounded and dying prey. Yet a walk in the woods in the light of day rarely reveals evidence of the carnage—or even animals that died of old age. They have been eaten by predators and scavengers, the detritus removed by beetles or eaten by other insects. "If it weren't for insects," says one entomologist, "we'd be up to our necks in dead bodies."[1] Nature cleans up after herself, concealing the fact of death.

Human civilization has its own methods. In 2008, 171 people died in my town. Most of them died in private, out of the eyesight of the public. Here's where 167 of them died:

64 in nursing homes
34 in hospices
32 in inpatient wards
27 at home
6 in the emergency room or outpatient clinics
3 were dead on arrival at the hospital
1 in the woods[2]

Only 4 people died in public, and all before a limited audience: 2 on the sidewalk, 1 in a dormitory room, and 1 was run over by a train on the tracks.

Most of us are spared the sight of a human actually dying except for the few relatives and friends we visit in seclusion at home or in an institution designed for secret dying, such as a hospital, hospice, or nursing home. As we walk the streets, people inside various buildings are discreetly dying, sparing us reminders of our own deaths. Americans associate September 11, 2001, when 2,700 people died in terrorist attacks, with death. In fact, it was an off year for human mortality. The Centers for Disease Control and Prevention (CDC) say that the usual 2.4 million Americans died, up about 14,000 from 2000, but the death *rate* actually declined.[3] The population expanded faster than it was being killed off. Seeing the towers crumble, however, it was easy to visualize the people inside and know that they were dying. We didn't have to imagine the jumpers. They died in plain sight. A video camera caught one young man, properly dressed in a shirt and tie—the coat having been discarded, of course—squatting on a ledge, preparing to jump. My mind pictured his refrigerator when he was a kid, all the straight-A report cards held by magnets. All those hoops jumped through to get the valued job in an investment banking firm or whatever. And it ended on a ledge. Americans reacted to the blunt images of death in a variety of ways.

I lived in Manhattan for twenty years and saw only one person die. On May 16, 1977, the right front landing gear on a Sikorsky S-61L helicopter collapsed as it landed on the roof of the Pan Am Building. One of the twenty-foot blades broke off and, after flying into the crowd waiting to board, killing four men, went over the side of the building and fell to the street below. I was walking on Forty-third Street when the rotor struck and killed a young woman who was walking a block ahead of me. I quit my job the next day.

Death today is primarily a secret affair, watched over by special, self-assured experts. It was not always so. Death was once everybody's business. In this chapter we'll look at how humans have pondered how and why people die and, most important, when they are finally, thoroughly, and decidedly dead.

We shall see the ongoing debate over where the pivotal source of life resides in the body. One can think of it as the "self," as did the medical personnel who declared Fernanda brain dead, but there is no need to be metaphysical. Simply put, is there an organ or region of the body that, if rendered inoperable, means that the person as a whole is dead?

The Egyptians and Greeks believed the heart was the important organ, but biblical Hebrews and Christians thought the breath was the key to life. For many centuries vying beliefs proclaimed the supremacy of one or both organs. The brain started to take on importance during the Enlightenment when doctors separated automated processes (breathing/circulation) from sensation and volition (based in the brain). Hence the move toward defining the brain as the important organ began. Another school of thought was more conservative, holding off on calling a person dead until putrefaction set in, the result, more or less, of total cell death. Following is a summary of this centuries-old investigation.

THE NATURE OF DEATH IN ANCIENT EGYPT

The ancient Egyptians believed the heart was the primary organ of life. They "reasoned that all the senses report directly to the heart, that it was the seat of thought, emotion, and intelligence," as well as the seat of conscience, which was to be weighed by judges in the afterlife.[4]

The high-ranking dead were embalmed to preserve their bodies for their entrance into the next world. The embalming process removed water, causing the body to dry and leaving little for decay-causing bacteria to live on. The major organs were removed, though the heart—associated with revitalizing the corpse—was often left behind. Male genitalia were propped up to give the appearance of virility (in one myth, the god Osiris engenders a son posthumously).[5] Medical knowledge limited the extent of mummification. Though the liver, lungs, and intestines were removed, the kidneys were left behind, most likely because ancient embalmers—who worked by touch—could not find them.[6] Medical historian J. Worth Estes suggests that the brain was removed, because "as far as the Egyptians knew, it had neither physiological nor philosophical functions."[7] We can surmise that the heart and genitalia were thought to be more closely related to life (and hence death) than was the brain, the opposite of today's thinking. The bodies were probably defleshed.[8]

Other hints about ancient Egyptian attitudes toward death come from vandalism of the tombs of the recently dead. A contradiction is set up: the very powerful and rich go through elaborate procedures to keep their memory alive, their bodies from decomposing, and their power intact, set against a culture of grave vandalism, robbery, and revenge. One owner included writing on his tomb "that cannot be erased." This and other texts suggest a continuing problem with vandalism by later robbers and by contemporaries who bore a grudge against the departed. Vandals might desecrate the tomb and erase the person's name in an attempt literally to wipe him out. Graffiti found next to a "mutilated figure" in an ancient tomb suggest that the rank and file perhaps believed the dead to be helpless, and the tombs, embalming, and elaborate ritual added up to not much.[9] The ruling classes of the ancient Egyptians (as well as the Inka) were in denial over death, constructing elaborate theoretical afterlives. The lower classes wrote doggerel on their tombs, mocking them and telling them that they were not coming back. For the poor and powerless, death was the great equalizer.

Ancient Egyptians watched the bodies of the newly deceased until they were sure they were dead before embalming them. Jacques-Jean Bruhier d'Ablaincourt, the translator of and contributor to an eighteenth-century text on methods of determining death, also suggests that embalming, which started with a cut into the stomach, was a final test to determine if the subject was truly dead. Despite the fact that embalmers were public officers, d'Ablaincourt recounts that they were customarily stoned by the public. The stonings could have been an expression of horror, but he decides it is more likely that they arose from mistakes made in prematurely declaring a person dead who then suddenly awoke under the embalmer's knife. If nothing else, embalming the person beforehand would certainly prevent premature burial, by definition.[10]

Peter Lacovara, the curator of ancient art at the Michael C. Carlos Museum at Emory University in Atlanta, Georgia, says the Egyptian tradition of mummification most likely started with the practice, from 5000 to 3000 B.C., of burying bodies in shallow pits in the desert of the Nile Valley. The bodies dried out and were preserved. "It was ingrained [in the ancient Egyptians], that the body must be preserved as a home for the spirit," says Lacovara.[11]

The tombs and pyramids that hold mummies were meant to simulate the effects of shallow graves in the desert. Lacovara says the ancient Egyptians were a superficial people; appearances were important. The mummy had to look, vaguely, like the person within, an idealized portrait in plaster. He also says that mummification was "more like taxidermy than embalming." Sawdust and packets of salt were used as fillers and preservatives. Wigs were applied, and artificial eyes were inserted. A reason the brain was scooped out and discarded, says Lacovara, is that the Egyptians noted that "the fish rots from the head." The disposition of the body was expensive in ancient Egypt and included the cost of mummification, tomb construction, the coffin, and food and other staples for the afterworld. It was a classic case of death denial. The temples became rich because wealthy Egyptians turned their landholdings over to them and in turn the temples oversaw their burials and preparation. To the ancient Egyptians, the afterlife was this life but better. It was a continuation, an "eternal vacation" called *shawabatis.* Drawings in tombs included big jars of perfume, furniture, food, little servant figures, big houses in the country.[12]

Lacovara compared the ancient Egyptians to Americans in the 1950s: "On top of the world. Insular and chauvinistic." There were advantages to their national pride. They were racially color-blind. "It didn't matter what color you were as long as you were from Egypt." The religion of ancient Egypt had no name. It wasn't considered religion. The Egyptians had beliefs, and they made no distinction, as we do, between religious beliefs and secular beliefs. The obsession with rituals surrounding death and a subsequent physical, concrete afterlife made for some interesting ramifications. They had a no-MIAs policy. Warriors killed in battle needed to be found and properly buried. Some were even buried in the king's tomb.[13]

In their efforts to extend their lives through mummification, the Egyptians became the deadest people on the planet. Emily Teeter, an associate curator at the Oriental Institute of the University of Chicago, in attempting to peer into mummies without tearing them apart, looking for signs of death and disease, had medical technicians scan a mummy using MRI. The techs were confused because nothing showed up on the screen. Nothing. They figured the MRI wasn't working, but to check, they tossed a bag of saline solution in with the mummy. The bag of salt water showed up on the scan, but not the body. The MRI was completely flat. "A successfully done mummy has no moisture. So it is dead, profoundly dead," says Teeter. The Egyptians loved life, she said, and they feared death. They worried about such things as how lonely they would be separated from their family. So, says Teeter, they tried to turn death into something they could understand. The soul of a mummy was represented by Ba, a bird that flew around during the day but returned at night to spread its wings over the mummy to preserve its personality. "The Egyptians never believed in reincarnation. A person's soul is his own, and he is always reborn as himself."[14]

DEATH IN CLASSICAL GREECE AND ROME

In classical Greece, Plato[15] and Hippocrates believed in immortality of the soul. In *Regimen* Hippocrates states, "No body perishes entirely, nor is anything made that was not before;

but mixed and separated things are changed about."[16] Meanwhile, Aristotle believed that the soul and body were inextricably united; when one died so did the other. Foreshadowing the loss of "personhood"—a modern definition of death—Aristotle defined the soul as "the vegetative soul of nourishment and growth and the animated soul of motion and sensation."[17] To him, death was a process of slow, irreversible loss of faculties. A person might lose his sight or lose his memory prior to the final loss, death. Life, by contrast, was a collection of possessions: the faculty of sight, taste, intelligence, strength—a concept that was taught for centuries in Europe.[18]

Democritus and Epicurus saw death as complete personal dissolution. With some analogies to current brain-death ideas, Epicurus considered death the moment when sensation ceased, corresponding to him with when one's "soul atoms [were] dispersed into the infinite void."[19]

Though their theories about death may have sounded pat, the Greeks and Romans were not complacent about declaring people dead. They were cautious. According to medical historians Mary Farrell and Daniel Levin, "Greek physicians believed death could begin in the brain or the lungs," but the heart was the determining organ. The presence or absence of a heartbeat indicated the presence or absence of life.[20] Nonetheless, both the Greeks and Romans knew that those signs were not dependable. The physician Galen described conditions that mimicked death, such as "hysteria, asphyxia, coma, and catalepsy."[21] Thus, the Greeks cut off a finger before cremation to ensure that death had taken place,[22] and Plato recommended delayed burials, ordering the "Bodies of the Dead to be kept till the third Day, in order to be satisfied of the Reality of the Death."[23] His *Republic* tells of an Armenian soldier who was left for dead. When his comrades returned to bury their comrades ten days later, they discovered all the bodies in a state of decomposition, with the exception of the Armenian, whom they carted home for a proper burial. The Armenian, however, recovered. The same incident (using a different nationality) appears in a later source, *Korman's Treatise de Miraculis Mortuorum,* also apparently drawn from Plato.[24] Korman's narrator, the German scholar Johannes Andreas Quenstedt (1617–1688), attested, "That the Soul sometimes remains in the body, when the Senses are so fettered, and as it were locked up, that it is hard to determine whether the Person is dead or alive."[25]

In Roman times, Pliny complained that doctors could not identify a dead person. The Romans knew that lack of breath, coldness, or corpselike appearance did not ensure death. Celsus (Aulus Cornelius, c. 25 B.C.–c. A.D. 50), the Roman cataloguer of medical knowledge, recounts concerns about premature burial. He writes:

> I know that if future Death is to be foretold by certain Signs, some may ask how it comes about, that Patients given over as dead by their Physicians, sometimes recover. . . . Medicine is a conjectural Art, and such is the Nature of Conjecture in general, that what succeeds most frequently may yet prove fallacious in particular Instances. It is therefore by no Means reasonable absolutely to divest that of Credibility, which hardly proves fallacious once in a thousand Times; since there is no Comparison between the Success and the Error. What I assert is not only applicable to the Signs of Death, but ought also to extend to Marks and Prognostics of the salutary Kind. . . . But it is certain that the Signs of Recovery or Death, are far more fallacious and defective in Acute, than in Chronical Disorders.[26]

In other words, it is better to err a thousand times in calling a dead person alive than once to declare a living person to be dead. In response, Romans put the body in warm baths or rubbed it with hot water in an attempt to revive it.[27] They also instituted delayed burials. For example, the Roman rhetorician Quintilian (Marcus Fabius Quintilianus (c. 35–c. 100) exhorted his fellows:

> For what purpose do ye image that long delay'd Interments were invented? Or on what Account is it that the mournful Pomp of funeral Solemnities is always interrupted by sorrowful Groans and piercing Cries? Why, for no other Reason, but because we have seen Persons return to Life, after they were about to be laid in the Grave as Dead.[28]

Conclamation, the tradition of calling the dying by their names in an attempt to revive them, was an important part of Roman funerals, probably predating Rome (and carried over into papal ritual). According to J. J. Bruhier d'Ablaincourt, the translator of Jacques-Bénigne Winslow's *The Uncertainty of the Signs of Death and the Danger of Precipitate Interments and Dissections:*

> Lanzoni, a Physician of Ferrara, informs us that when any Person died among the Romans, his nearest Relations clos'd his Eyes and Mouth; and when they saw him ready to expire, they catched his last Words and Sighs. Then calling him aloud three times by his Name, they bid him an eternal Adieu.[29]

Ovid also refers to the loud calling repeated during the funeral after the dead person's eyes were closed. This is attributed to making sure the person was really dead, though the action became part of religious ritual.[30] The custom of trying to catch the last words and breath continued into the Christian era (see "Early Christian Thinking," below).

HEBREW DEATH

Though early sources do not precisely define the moment of death, definitions can be inferred from the Bible and the Talmud. Talmudic scholars decreed that if a man was buried, one had to dig up to the nose in order to determine if he was dead, because "the main sign of life is in the nose, as it is written: 'All in whose nostrils is the breath of the spirit of life'" (Genesis 7:22).[31] Hebrew scripture also describes how the angel of death stands over the dying and drops a drop of gall into the mouth, which kills the man. His face "turns greenish [*morekoth*, like the color of gall] and his body develops a bad odor."[32] The greenish face and odor are signs of death.

But the conclusive sign of death is when "the golden bowl is shattered" (Ecclesiastes 12:5–7), a biblical reference to the abdominal swelling in a corpse. Before this swelling occurs, watchers are forbidden to move the body or even touch it, lest the tiniest disturbance be what kills the person. According to Avraham Steinberg, who compiled an encyclopedia of Jewish medical ethics, "Rabbi Meir used to explain this matter by an analogy: he who touches a flickering lamp extinguishes it."[33] But the only truly conclusive sign of death is putrefaction. The dead were put on sand or salt to preserve them, but the survivors would go and examine the dead for three days to ensure that death had actually occurred. (See twentieth-century Africa for similar customs).[34]

Early Jews also thought that cutting the Achilles tendon

would automatically result in death. See Psalms: "Evildoers lie in wait for my heels, as if they watch for an opportunity to kill me."[35] But legal scholars insisted that even in a man with a severed tendon, death was certified through the "testimony of witnesses," not by the nature of the injury.[36] So in putting together legal definitions of death, Hebraic scholars would make a distinction based on likelihoods. The Talmudists say that if a fisherman disappears underwater and later one of his feet is found, the man is not legally dead, because he might have survived. However, if the whole leg with the foot is found, the man is declared legally dead.[37]

Later some Hebraic schools declared lack of respiration to be the definitive sign of death, while others described death as lack of heartbeat. The debate went back and forth. Some rabbis believed the breath was only one sign of life, to be used in combination with heartbeat and pulse. Other rabbis held that breathing was the primary sign of life, though if breath were absent but other vital signs were present, the person was still alive. Others maintained that *irreversible* lack of breath always indicated death.[38] A stopped heart as an indicator of death came later in Talmudic thinking, but not for modern reasons. Medical theory at the time held the heart to be part of the respiratory system, sending warm air (pneuma) to the lungs.[39]

SLAVIC DEATH

D'Ablaincourt recorded burial customs of early Slavic people (circa sixth century), who rubbed the dead body with warm water for an hour. The body was then publicly displayed for two days, and on the third day it was buried, accompanied by loud grieving and crying. According to d'Ablaincourt, "Singers spent an Hour in shouting and making a Noise, in order to recall it to life; after which it was let down into the Grave. . . . So that this People used the Test of warm Water, that of Cries, and a reasonable Delay, before they proceeded to the Interment."[40]

EARLY CHRISTIAN THINKING

Christian writers in the second century feared spiritual death more than physical death. St. Ambrose (c. 339–397) considered death to be merely the passage of the soul to Heaven—nothing to fear unless your soul was corrupted. Echoing Plato, he said that the physical body was the "temple of the Holy Spirit," the vessel holding the essence of life, but not life itself.[41] Despite those teachings of the Church, the general populace apparently wasn't convinced of immortality. During a plague, St. Cyprian (c. 200–258), bishop of Carthage, chastised the city's inhabitants for their attempts to heal themselves: "What madness it is to love the afflictions, and punishments, and tears of the world and not rather to hurry to the joy which can never be taken from us." St. Augustine also commented on the paradox when the sick implored God to stop their suffering and let them die, while simultaneously hiring a physician at any price to find a cure.[42]

Early Christians believed the resurrection entailed both the person's spirit as well as the full physical body as in life. The American cultural anthropologist Bilinda Straight, paraphrasing the twentieth-century scholar Fernando Vidal, claims that starting with St. Paul and continuing into the twentieth century, Christian writings asserted that humans exist only in bodies. The soul is part and parcel of the body and cannot be separated from it. Thus, how God would manage to resurrect each individual in his or her full human body was a dilemma and fueled debates for centuries.[43]

The hands and feet of the dying person were stretched out

to help the soul escape. Dying was gradual, the ancients believed. The soul left moving from the feet up through the body and exited out the mouth. This reflects the progress of some diseases in which various body parts grow cold prior to death.[44] The person was also given Communion, and as he expired those attending would try to catch his last breath in their mouths. According to Rush, this last breath was believed to be the departing spirit. Wives, husbands, and parents tried to take in some part of their beloved's soul and also to give their own life to the dying person.[45]

MEDIEVAL EUROPE

The twelfth-century Jewish philosopher-doctor Maimonides in Islamic Spain believed the head, not the breath or heart, gave the soul its central guidance. He noted that a body without a head might move but the movement was spasmodic, without direction. Therefore he was an early voice in designating the head/brain as the important organ.[46] Meanwhile, most other medieval scholars believed that either the heart or the lungs, or both, were the primary organs that maintained life.

From 1200 onward Christianity began to focus on the physical body. As summarized by Lawrence Conrad, a historian of near-eastern medicine, et al., Christ was physical and therefore his resurrection was physical. The body parts of saints, whether a hair clipping or a finger, had the power of life and took on enormous power. Christ's body could be reexperienced in the communal wafer, which miraculously turned into his flesh in the worshipper's mouth. In fact, in the twelfth century, Bishop Hugh of Lincoln bit into an arm-bone relic of Mary Magdalene; in his defense he reasoned that if he could eat the body of Christ, why couldn't he eat the arm of a saint?[47]

Thus religious scholars believed the body would be reformed in full after death, regardless of the manner of demise.[48] How to restore bodies that had been eaten or otherwise destroyed? This was discussed at length in the thirteenth century in "reanimation" theories. Medieval universities, in contrast to the theologians, looked to Aristotle's theory that humans were made of both abstract forms and physical matter.[49] St. Thomas Aquinas explained that reanimation followed Aristotle, stating, "Soul is the 'form' of a natural body that potentially has life." Once body and soul are reconnected, life reappears.[50] Yet according to Bilinda Straight, "The doctrine of the resurrection went against the grain of specific cultural understandings from at least A.D. 200 and onward through the medieval period." She implies that the cultural understanding was that physical death was final, despite all the teachings of the Church.[51]

The literature of the time reveals that people remained uncertain about who was dead and who was still alive. D'Ablaincourt compiled more than 150 pages of accounts of premature burial and mistaken death in his appendix to Winslow's book ("Additions"). The stories range from ancient times to his own era in the mid-1700s and make obvious that death has been uncertain for centuries. Its most commonly used signs in medieval times were stiffness, coldness, and lack of breath, heartbeat or pulse, and movement. When people were resuscitated, it was, at least officially, often attributed to miracles rather than poor methods of determining death.

In the year 1303, three accounts of resurrection were attributed to St. Thomas de Cantilupe. All three accounts were most likely cases of premature declaration of death and shed light on methods of determining death in medieval times. (Thomas apparently caused a lot of resurrections and got to be a saint as a result; see Robert Barlett's *The Hanged Man,* discussed in

Straight.)[52] In one account, a two-year-old boy fell into a castle moat. According to Straight, "Since his parents were at an all-night vigil for a dead neighbor, they did not know of Roger's fall until the next morning." At that point, the child was "stiff as wood and cold as stone." When the coroner arrived to record the death, someone noticed that the boy's tongue had moved. His mother put her tongue into his mouth to see if he was breathing and could sense nothing. He was completely cold, but she warmed him against her skin, and eventually he revived.[53] By this account we can surmise that breathing was considered a primary sign of life and the boy's stiffness and coldness were signs of death. Note also the vigil for the dead neighbor: vigils throughout history have functioned to ensure that the dead were really dead.

CAMERLENGO

Other clues to how medieval people determined death are found in the office of the papal camerlengo. This post was established sometime between the twelfth and fourteenth centuries with the office holder's duty expressly to ensure that the pope was indeed dead before burying him.[54] Following the ancient Roman custom of conclamation, the camerlengo calls out the pope's Christian name three times. "The silence of the dead bears the answer that is sought," declares Greg Tobin in his book, *Selecting the Pope: Uncovering the Mysteries of Papal Elections,* which bears the imprimatur of Monsignor Robert J. Wister, a professor of Church history at Seton Hall University. According to him and other sources, the camerlengo then taps the pontiff's forehead with a silver hammer or drapes it with a cloth.[55] The use of the hammer is unsubstantiated. Perhaps it is meant as a means of provoking a pain reflex. The camerlengo's colorful history and strange duties are entertaining to research and read about but difficult to pin down.[56] Were I a pope or a person of similar importance, I would very much like to employ a camerlengo. When he declares the pope to be dead, the camerlengo's job ends. It is a great job with few duties, so it seems unlikely that he would fake his boss's death.

RENAISSANCE EUROPE

Today we expect doctors to be at the bedside of a dying person, but in medieval and Renaissance times, doctors were noticeably absent. Indeed, doctors who accepted money from someone who was sure to die were viewed as unethical. Doctors were expected to tend curable ailments, not hopeless ones, and dying was a cultural and religious affair, not a medical one. The dying person was surrounded by family and clergy, preparing to cross the boundary from mortal life to Heaven. A whole body of writings existed, called *Ars moriendi, Art of Dying,* to instruct people how to face the end. People were expected to compose themselves, be calm, forgive their enemies, and resolve their conflicts with others. Catholics received the last rites, and all Christians believed that welcoming the coming transformation was necessary; otherwise the Devil would have more of a chance to make off with their souls.[57] In this religious and cultural environment, *how* one died was critically important. In England, churches paid people to watch over the poor during their last moments. In Italy hospitals made sure a nurse was present at all times to read religious texts to the dying.[58]

During the sixteenth century, scientific and logical reasoning started to compete with traditional religious views of the universe. In the 1500s, medical conventions followed Galenic medicine and Arabic medieval physicians, particularly Avicenna, the tenth-century Islamic scientist and polymath, whose teachings dominated medieval Europe and who had argued that aging involved the drying and loss of warmth in the body. Life consumes both moisture and heat; each person is endowed with a certain temperature and moisture, which is slowly consumed.[59] Death occurred once the body contained only water and shit, which could not be burned for heat by the body.[60] But in this context, many medical scholars voiced new ideas. Girolamo Cardano, a sixteenth-century Italian scholar known for his medical writings, reduced life and death to a mixing and separating of elements.[61]

By the 1600s reason gradually replaced much of the religious thinking on how the universe functioned. God set the universe in motion according to set laws, and the body was no exception.[62] Georg Ernst Stahl (1660–1734) suggested that the decay of the body was due purely to physical causes.[63] The century was also a time of experimentation. The first case of "brain death" in an organism maintained on a "respirator" may have been described by William Harvey, who preserved the circulation and respiration of a decapitated rooster in 1627.[64]

Both scholars and medical people struggled with seventeenth-century concepts of resurrection. According to Straight, one solution was to posit that a tiny body or personhood was contained in a "stamen," which then grew to become the actual person. Straight later describes debates on the location of memory and the role of the brain in personality:

> As the Enlightenment continued, other writers would debate whether memory could be contained in one "organ" or brain area or was dispersed throughout the

> nervous system—the decentralized view. . . . The decentralized view can be intriguingly paralleled incidentally, with contemporary challenges to mind/body dualism in cognitive neuroscience. . . . The stamen or embryological view likewise has contemporary parallels in understandings and uses of DNA that in some ways do indeed seem to mimic reanimation.[65]

If Shakespeare can be believed, Renaissance people determined death by testing for breath by using a mirror or feather. In Shakespeare's *King Lear* (written c. 1603), Lear bemoans the death of Cordelia, exclaiming:

> She's gone forever!
> I know when one is dead and when one lives;
> She's dead as earth. Lend me a looking-glass;
> If that her breath will mist or stain the stone,
> Why, then she lives. . . .
>
> [If] This feather stirs; she lives[66]

People were well aware of the difficulties in relying on any one sign of death (e.g., lack of breath or pulse, stiffness, coldness). A correspondence between the surgeon William Fabri and a physician at Neufchâtel, John James Crafft, is described by d'Ablaincourt, who came across it in Simon Goulart's 1628 *Admirable and Memorable Histories Containing the Wonders of Our Time.* Fabri records three incidents of premature burials during the plague years, an easy error when quick burials were prompted by the need to contain the disease. To paraphrase one story: a man in 1566 succumbed to the plague and after four days was believed to be dead. Eight hours after he was declared dead, "the Persons who were putting him in his Coffin, finding that he was neither cold nor stiff, began attentively to examine the State of the Body, in which they still perceiv'd a small and languid Degree of Respiration."[67] Yet he did not have rigor mortis or lose heat. Hot bricks on the feet and "a small quantity of Malmsey Wine," brought him back to life.[68] Dr. Crafft responded with five of his own accounts of premature interments. In one, a man in Neufchâtel was "seized with a violent Cardialgia, fell into so profound a Syncope [loss of consciousness], that he was taken for dead." Dr. Crafft thought otherwise. He blew pepper into the man's nose, which succeeded in rousing him.[69]

Throughout the centuries decomposition was the only sure litmus test, and this era was no different. Paolo Zacchia (1584–1659) was the physician of Pope Innocent X and noted for contributions to legal medicine. Jacques-Bénigne Winslow, writing in 1746, quotes him: "That there is no other infallible Proof of Death, but a beginning Putrefaction in the Body."[70] Terilli, a doctor in Venice, agreed and is quoted in Winslow as saying

> Since the Body is sometimes so depriv'd of every vital Function, and the Principle of Life reduc'd so low, that it cannot be distinguish'd from Death, the Laws both of natural Compassion and reveal'd Religion oblige us to wait a sufficient Time for Life's manifesting itself by the usual Signs. . . . Now the Time . . . allotted comprehends three natural Days; and if during this Interval no Marks of Life should appear, but on the contrary the body should diffuse a fetid and cadaverous smell, we may rest satisfied with the Certainty of the Death.[71]

ANATOMY THEATERS OF THE RENAISSANCE

During the Renaissance, anatomy theaters were built, primarily in the form of amphitheaters, that revealed more about life and death, and perhaps the shady area in between, than many people could handle. They were festive affairs, with showmen-anatomists performing before audiences, dissecting freshly dead bodies, holding up organs and other body parts to the delight of the spectators. Sometimes there was an orchestra. Tickets were sold, and attendees showed up wearing costumes.

Anatomy theaters began springing up in the early 1500s, primarily in Italy, and continued into the seventeenth and even the eighteenth centuries. They were located, among other places, in Pisa, Rome, Ferrara, Turin, Venice, and Mantua, with a spectacular theater in Bologna known for its opulence. The Germans and Dutch also had anatomy theaters. Anatomists who became stars included Andreas Vesalius, Bartolomeo da Montagnana, Antonio Benivieni, Matteo Realdo Colombo, Costanzo Varolio, Gaspare Tagliacozzi, and Alessandro Benedetti. Even François Rabelais, the French writer who was also a doctor, is said to have performed a dissection. Vesalius, who normally appeared onstage in Padua, made a guest appearance in Bologna's theater—the Carnegie Hall of the anatomy circuit—and there dissected three human corpses and six live dogs. The public dissections were never done in summer, only in winter, because of the smell.

With so many theaters and anatomists, bodies to dissect were at a premium. Generally, executed criminals were used. Dissection was considered punitive at the time, meant to extend punishment beyond execution. In fact, in perhaps a conflict of interest, some anatomists moonlighted as executioners. Other bodies used in the theaters were foreigners or out-of-towners—those living at least thirty miles away. Hospital patients were also candidates. Essentially, the anatomists were looking for anyone likely not to have relatives nearby who would protect the body.

As we'll see in chapter 7, Washington, D.C., has a similar law regarding organ donation. If a family doesn't contact a hospital immediately, its loved one will be "preharvested," dissected, and filled with preservative. In one case in Bologna, the authorities, under pressure from the anatomists, "reverse commuted" one criminal's sentence from life to death. His body was needed for the stage.

The theater performances were begun for medical students, but they soon became social events, and they were thought to have socially redeeming value. Benedetti said that public dissection revealed "the mysterious profundities of nature." Upwardly mobile professionals might sponsor a dissection for their friends and business colleagues as a means of self-promotion.

As for the dissections themselves, they sound a little scary. Doctors of the era agreed that even after a person dies, his body retains a degree of "sensitivity." Sometimes it was more than that. Vesalius complained that it was difficult to observe moisture in the cardiac membranes as it dissipates quickly after death. Finally he procured a man who had just been killed in an accident and, "eager to see this water," took out what he described as "the still pulsing heart." The anatomists learned, perhaps, that death does not come as quickly as we believe. One, Niccolò Massa, asked to be left unburied for two days "to avoid any mistake."[72]

EIGHTEENTH-CENTURY EUROPE

The eighteenth century was increasingly dominated by a "scientific" outlook on dying and death.[73] Scholars such as Albrecht von Haller (1708–1777), a Swiss biologist, conceived of decay as the result of unstable unions of bodily elements that deteriorate.

Both in common thought and in academia, life was increasingly seen as an anomaly in nature, rather than as its highest conception (as per Aristotle). Meanwhile, death and the dissolution of the body were perceived as following natural law.[74] Thus began, as one researcher put it, the "medicalization of death."[75] Doctors began to appear at the bedside of the dying to administer opiates. The Christian traditions of welcoming the transformation into Heaven and fighting off the Devil at the moment of death declined. Because of that, dying people no longer needed to be in their full faculties to prepare themselves for the transition. Instead they were now drugged to alleviate pain. This is a bit of a paradox. In medieval and Renaissance times, the dying needed to muster all their rational faculties to meet death head-on. But with the advent of modern medicines and a scientific view of death and the body, the patient himself no longer needed to be rational or in his faculties.[76]

As the boundary between life and death became more confused, increasingly medical technologies were introduced to tell the difference. In addition, the eighteenth century saw a marked increase in the morbid fear of premature burial, and the moment of death was intensely scrutinized. According to Farrell and Levin, artificial respiration appeared in France by the mid-1700s and was widely adopted by 1767 to help "drowning and suffocation victims."[77] Farrell and Levin also documented the use of smelling salts (1721) and electric shock (1755). Electric shock was even used to resuscitate a person, later becoming the basis of the creation of Frankenstein's monster.[78]

There were many, many accounts of premature decrees of death and burial and many quasi-scientific inquiries into same. Benjamin Franklin declared in 1773, "It appears that the doctrines of life and death in general are yet but little understood." He then cites an experiment in which flies preserved in Madeira wine were resurrected after surviving the journey from the bottler in Virginia to their destination in England. Franklin and friends revived two of the three by straining the liquor off the insects and putting them into sunlight.[79] Earlier that century more confusion between the living and dead was created by discoveries of bacteria surviving for months without water and worms living for decades in a suspended state.[80]

In addition, medical schools increasingly required dissection in their anatomy programs, despite laws that outlawed autopsies except on executed criminals. Indeed, a librarian of the British Royal College of Surgeons recorded that before bodies for dissection were plentiful, "a surgeon might be punished in one Court for want of skill, and in another court the same individual might also be punished for trying to obtain that skill."[81]

But the gradual acceptance of autopsies both expanded doctors' understanding of what caused people's deaths and increased the public's fear of being buried prematurely (because autopsies sometimes demonstrated that "dead" people were alive) or becoming fodder for the anatomy theaters. In the sixteenth century, King Henry VIII had decreed that criminals could be autopsied, resulting in an active trade for their bodies.[82] From the early eighteenth century to the nineteenth century, bodies for British medical schools were provided by "resurrectionists" (body snatchers), who dug up freshly buried bodies at night and delivered them to the schools for an exorbitant fee.[83]

During that time, government decrees prohibiting too-speedy burials came on the heels of public fears and the increasing confusion of scholars as to just what constituted death. The eighteenth-century Italian physician Giovanni Maria Lancisi, commenting on the classical Roman's Quintilian's exhortation that living people had too often been put into the grave as dead, wrote, "For this Reason the Legislature has wisely and prudently prohibited the immediate, or even too speedy Interment of all dead Persons; and especially of such as have the Misfortune to be cut off by a sudden Death."[84]

Winslow also cites a similar decree set forth in 1772 by the Duke of Mecklenburg (a region now part of northern Germany) to prohibit "immediate burial and requiring a three-day wait after the establishment of cardiopulmonary death by physicians before burial could occur. The signs of death required by this decree were the appearance of *livedo reticularis* and the putrefaction of the flesh."[85] (The Mayo Clinic defines *livedo reticularis* as a "vascular condition characterized by a purplish mottled discoloration of the skin."[86]) Jewish law, which decreed that the dead should not lie unburied overnight (somewhere this had changed from the three-day custom in biblical times), was largely overruled by the Mecklenburg decree, which was adopted by many other countries as well.[87]

Signs of Pulse and Respiration

Jacques-Bénigne Winslow, writing at mid-century, recorded many of the difficulties and techniques used in his time. In his book *The Uncertainty of the Signs of Death and the Danger of Precipitate Interments and Dissections,*[88] he laid out numerous errors and signs used to determine demise: paleness, coldness, rigid limbs, and the loss of all senses. Yet all of these, he said, were not certain signs of death. His own interest came because of having himself been twice pronounced dead by physicians. While asserting that the pulse and signs of breathing are the true signs of life, he acknowledged that they can sometimes be invisible, even to experienced practitioners. Patience, said Winslow, was essential to finding a faint pulse, since the hand must be carefully positioned to permit the easiest passage of blood through the veins.

Conversely, pulsing can be felt at the fingertips in people already dead. Therefore he advised skilled practitioners to seek out the pulse in other areas: the temples, carotid arteries, or near the heart.[89] To these tests mentioned by Winslow, Farrell and Levin add several others: cutting open arteries and checking for "livid spots, pallid skin, depressed limbs, and sunken eyeballs."[90]

Patience was also needed to ascertain respiration in subjects whose breath was "languid" and "overpowered." Indeed, Winslow asserted that the breath can continue simply by the forces within the lung structures, helped by the circulating blood, however faint. Because respiration is so hard to detect under such circumstances, a number of other methods were developed. They included holding a candle or cloth up to the nostrils to see if it wavered, a method he believed was prone to great error. The mirror test was also unreliable according to Winslow, who asserted that gases from decay can seep out of the nose and mouth of a corpse. Another method places a glass of water directly below the breastbone.[91] All those methods were prone to error should the attendant's hand move or some other motion influence the flame, cloth, or water.

To those tests, Farrell and Levin add using a feather, looking for bubbles in a submerged body, and testing for breath by using a hygrometer.[92]

Stimulating Pain and Reactions

Growing in popularity as a means of finding life were methods that elicited pain or other strong reactions. Winslow lists several resuscitation methods that stimulated the senses. They included putting stimulating or strong-smelling items near the nose; rubbing the gums with salts, liquor, garlic, horseradish, sharp objects, etc.; stimulating sensitive areas of the skin (fingertips, toes) with sharp or stinging objects ("whips and nettles"); injecting smoke and other irritants into the intestines; and violently jerking the arms and legs.[93]

In the eighteenth century some doctors knew that people rescued from nearly drowning could appear to be dead. In 1774 in London, the Royal Humane Society was founded by two doctors, William Hawes and Thomas Cogan, to help resuscitate those who had drowned. They taught resuscitation techniques and offered rewards for saving lives.[94] The first award was given in 1774 to Thomas Vincent, who pulled a year-old toddler from an aqueduct. The account tells that seven to ten minutes had passed by the time the child was retrieved. According to the annual report of the society: "The women upon the strictest examination affirmed, that the child was to all appearance dead; its eyes were fixed, it lay breathless, and void either of motion or pulse." The boy was resuscitated through a combination of shaking, beating, and having his belly and chest rubbed with salt.[95]

In addition to shaking and rubbing, loud noises (e.g., from a trumpet) and yelling into the patient's ears were used, mimicking procedures in other cultures and other times. In Winslow's time, the sense of hearing was thought to be the last to fail in a dying person, since people who recover could often recount conversations but nothing else.[96] In 1788, Charles Kite introduced a test using electric stimuli to invoke twitching, which Farrell and Levin called "the distant ancestor of the flat electroencephalogram (EEG)."[97] Other methods included applying caustic chemicals to see if the skin would blister and using "artificial ventilation." Failure to respond to either method indicated death.[98]

If such measures did not work, more severe measures were taken to stimulate the senses or cause pain—anything to wake the person from his condition. According to Winslow, body parts were burned, skinned, lacerated, or cut to cause pain. Needles in the palms of the hands or soles of the feet and cutting of the shoulders and arms were also used. Winslow cites an apoplectic woman destined to be buried who was saved by a needle thrust under her toenail. Burning was thought to be the most effective of all; thus red-hot irons were applied to the feet or the top of the head. When one physician noticed that a patient with no pulse and no breath still had flexible limbs, he rubbed the patient's feet violently with a hair cloth saturated in salt water, and in less than an hour the person revived. According to Winslow, if all these labors did not revive the person, he was probably dead.[99]

Nascent Brain Death

It is in the eighteenth century that we begin to see a shift from what today we would call cardiopulmonary death to brain death. The use of stimulation of the faux dead ties in with the work of William Cullen, a Scottish doctor. Cullen believed that "nervous sensibility and muscular irritability were the principles of life," rather than respiration or circulation. Farrell and Levin see his decision to place life as coming from the "sensorimotor" faculties rather than from cardiopulmonary processes as paving the way for ideas of brain death.[100] They also identified other medical thinkers who laid the groundwork. Marie François Xavier Bichat, a French anatomist and physiologist, for example, differentiated between "organic life," the heart, lung, and other functions, and "animal life," or the sensation and volition of the brain. He stated, somewhat mysteriously, "The patient may live internally for several days after he has ceased to exist beyond himself."[101]

Later, in 1834, the physician A. P. W. Philip wrote "On the

Nature of Death," a philosophical paper in which he defines death as the loss of sensation and human will. In addition, he cites experiments that showed the faculties of sensation and will to be located in the brain:

> The sensorial functions constitute the sensitive system—that by which we perceive and act. . . . The nervous and muscular, the vital system, that by which we are maintained. From the same experiments it appears, that what is called death consists in the loss of the first of these classes of functions, the sensorial, the nervous and muscular functions still continuing, which are only lost as a consequence of the failure of respiration.[102]

With the casting of will and sensation as the seat of life, so, Farrell and Levin assert, begins the movement for brain death as defining the end of life.[103]

NINETEENTH-CENTURY EUROPE

By the nineteenth century, the concept of death underwent significant changes due to increased autopsies, which now became routine. The continent's medical schools once again included surgery with studies of internal medicine after centuries of social strictures that held it to be unethical for a doctor to physically examine a patient or use tools in the diagnosis or cure. Study of corpses revealed the connections between the body's internal structures and its state of health and cause of death.[104] On the heels of this change, medical innovations such as the stethoscope, in 1816, followed. Autopsies led doctors to conclude that the body's anatomy could be explained only in terms of survival, that is, the organism escaping from death. Thus life and death were seen as entwined and necessary to each other.[105]

During the nineteenth century, the boundaries between life and death got messier. Research on hibernating animals revealed that their breathing and blood flow slowed almost to a stop, leading to questions of what exactly life was. Indeed a doctor at the time, Dr. Paul Brouardel, discovered that a hibernating animal could be cut and hardly bleed. He also experimented with trying to stimulate the heart and other muscles in hibernating animals: "Tap the heart with the end of your scalpel, and you will not induce a contraction, and it will be equally impossible to arouse muscular contractibility." When the temperature warmed up, however, the apparently "dead" creatures came back to life.[106]

As a result, the debate over determining death intensified as the general public became concerned about being buried alive. Current periodicals carried tales of horrible mistakes, including an 1824 first-person account—of questionable veracity—of a man rescued by the resurrectionists in the nick of time (see endnote).[107]

A member of the Royal Academy of Medical Sciences in London, William Tebb, collaborating with Edward Perry Vollum, a U.S. Army medical inspector, wrote an entire book, published in 1905, on preventing premature burial that provides vast detail about generally accepted signs of death: lack of pulse, breath, or movement and the onset of rigor mortis. Yet all those methods had pitfalls; doctors and other researchers increasingly relied on more sophisticated techniques as medical knowledge increased, a trend that placed death determination in the hands of professionals. Much of the following section is drawn from Tebb or cited by him.

Pulse and Breath

During Tebb's time, popular belief held that lack of pulse and breathing indicated death. However, like Winslow in the previous century, Tebb could recount numerous instances of people lacking both, and then reviving. He tells of a woman who suffered from "death spells" during which she had no perceptible heartbeat, pulse, or breath, only to revive in two to four days.[108] In another case, a doctor using a stethoscope on a woman, "failed to detect the faintest pulsation in the heart," yet later the patient revived.[109] Such confusion fueled debate among physicians. One doctor testified that if the heart stopped for two minutes, the person was dead, then later declared that one needs to listen for at least half an hour. Another doctor declared that it was impossible to listen to the heart that long, because all kinds of noises are heard, including the sound of the physician's own heart.[110]

As I mentioned before, popular belief in Tebb's time (as well as in earlier centuries) held that death could be determined by holding a mirror in front of the mouth. Yet Tebb cites Sir Benjamin Ward Richardson, the author of "The Absolute Signs and Proofs of Death" (in *Asclepiad,* 1889) and a contemporary of Tebb, specializing in "proofs of death," as having discredited this test. Richardson speaks of "life at low tension," by which he probably meant suspended animation or coma, along with the effects of chloral poisoning and catalepsy, in which "even the most practiced eye is apt to be deceived."[111]

Movement

Popular belief held that death was manifested when movement ceased for a long period, especially if the jaw dropped. But, Tebb points out, this was not dependable because catatonic patients can be immobile and the jaw does not always drop at death (e.g., after strychnine poisoning). In addition, shortly after death, movement can be stimulated by electrical impulses.[112]

Rigor Mortis

During rigor mortis a cadaver can become so stiff that it can be lifted like a suitcase. Tebb, however, noted that rigor mortis varied: sometimes it occurred immediately; at other times it was delayed, especially in people in good health; sometimes it did not occur at all. It also could occur in living patients with certain conditions. Thus Tebb cites a Dr. Roger S. Chew, who said, "Rigor mortis is a condition that seldom or never supervenes in the hot weather in India, and is often a feature of catalepsy."[113]

Circulation

A common test of the time was the diaphanous test: the hand was examined with fingers extended and touching one another, backlit by a strong light. If red lines appeared between the fingers, the person had some circulation and was alive. If not, he was dead. The French Academy awarded this method a prize for determination of death, but Tebb discredited it because one patient who had proved to be dead by eight other tests still revealed red lines under the diaphanous test, and another patient whose fingers showed no red lines at all had merely fainted and was shortly thereafter revived.[114]

Decomposition

According to Tebb—in concert with a long line of careful medical observers before him—the onset of decay was the only reliable proof of death. He quotes numerous doctors of his time to this effect.[115] A French medical writer in an article on real versus apparent death declared:

> Experience has shown the insufficiency of each of these signs, with one exception—putrefaction. The absence of respiration and circulation, the absence of contractility and sensibility, general loss of heat, the hippocratic face, the cold sweat spreading over the body, cadaveric discolouration, relaxation of the sphincters, loss of elasticity, the flattening of the soft parts on which the body rests, the softness and flaccidity of the eyes, the opacity of the fingers, cadaveric rigidity, the expulsion of alimentary substances from the mouth—all these signs combined or isolated may present themselves in an individual suffering only from apparent death.[116]

The aforementioned Dr. Richardson, the expert of the day in death determination, in "The Absolute Signs and Proofs of Death,"[117] listed eleven signs of death, including respiratory failure, cardiac failure, rigor mortis, and others. The eleventh sign was "putrefactive decomposition" (see endnote for complete list).[118] Richardson declared that if the first ten signs pointed conclusively to death, then death was actual. Only if there was doubt did the physician have to wait for decomposition. As for himself, Richardson instructed his family not to bury him until clear signs of decomposition had occurred.

New Technology

Gadgets invented in the nineteenth century allowed physicians to examine death in more sophisticated and microscopic ways. To Richardson's list, Farrell and Levin add a number of other methods: checking vital signs using a stethoscope, electrical tests, and body temperature.

> [M]icrophones amplified chest sounds, radiographic fluoroscopy searched for the motion of vital organs, and the ophthalmoscope was used to examine the circulation of the retina. Time-lapse photography could record the reaction of pupillary reflex irritability.[119]

These and other signs of death received varying degrees of official acceptance. In 1892, the state of Wurtemberg, in what is now southern Germany, set out guidelines for issuing death certificates, which emphasized testing for a lack of reflexes or response to pain.[120] What is notable here is the lesser importance of the breath and heartbeat compared to the more primary importance of "sensibility" with its connection to the brain—and circulation.

In 1885 the *British Medical Journal* published "Death or Coma?," an article discussing the difficulty of using any one sign of death as proof other than the fail-safe putrefaction. However, the article declared that true death was attended by several signs, all of which should be present after five days. The signs are: sunken eyes, "flaccid cornea, and dilated insensitive pupils," and absence of heartbeat and breathing as determined by using a stethoscope. The article questioned using rigor mortis as a standard at all, and instead recommended "post-mortem lividity of dependent parts" [I assume this means livor mortis][121] Thus we are down to four signs: the eyes, breath, heart, and livor mortis.

We see a further "medicalization" of death determination and the attendant infallibility of doctors. Ten years later, the same journal ran an article that labeled contemporary fears over premature burial as "hysterical outpourings." According to the *British Medical Journal,* laypeople were not qualified to determine death, which now was in the hands of doctors:

> The possibility of apparent death being taken for real death can only be admitted when the decision of the reality of death is left to ignorant persons. We are quite unprepared to admit the possibility of such a mistake occurring in this country to a medical practitioner armed with the methods for the recognition of death that modern science has placed at his disposal. . . . During [the period prior to putrefaction] various signs of death appear which, taken collectively, allow of an absolute opinion as to the reality of death being given.[122]

Tebb, for his part, sided with numerous medical authorities who questioned the *Journal*'s assertion that an "absolute opinion" was possible.[123]

NON-WESTERN TRADITIONS

Non-European traditions surrounding death determination are as rich as our Western traditions, simply not as well documented. Here are some highlights from various cultures.

The Americas

Natives of Florida were burned by a fire once dead, first one side, then the other. This custom was described by the French humanist Muretus (Marc-Antoine Muret, 1526–1585) and interpreted by d'Ablaincourt as, most likely, a way to prevent premature interment.[124]

In the Caribbean, people lamented over the dead person and then addressed him at great length about all the things in life he once loved, entreating him to come back to life with a repeated refrain: "How comes it then that thou hast died?" After these entreaties they placed the body in an uncovered grave and offered it food, still hoping that it might yet eat. After ten days of no eating, the grave was finally covered, a sufficient time having elapsed that a revival would have occurred if it was going to.[125]

D'Ablaincourt also recorded that "Americans" mourn extensively for a child or young adult, only moderately for a middle-aged person, and "so transitory as hardly to testify the smallest Degree of Sorrow" for an old person. But he did not specify whether he was writing about native Americans or colonists.

Asia

Chinese tradition also included making loud sounds at the point of death. By the eighteenth century this was simply a tradition, but it also functioned in the same way as the Roman conclamation. At the moment of death, Chinese relatives broke out in a loud "death howl." In ancient China it is recorded that people stamped in addition to the wailing. According to the sinologist Jan Jakob Maria de Groot, the wailing was meant to try to bring back the dead. A meal was laid out for the corpse.[126] Both ancient and eighteenth-century Chinese then went through a period of watching the corpse for a number of days; relatives would sleep next to it until burial.[127] Right before burial, de Groot describes a final attempt to rouse the person, in which the eldest son cries out, "Father (or mother) stand up!"[128]

In ancient Persia, the dead were left out for animals to devour. According to d'Ablaincourt, Persians put out their dead without waiting for putrefaction. If the body was eaten quickly, it was a good omen. If the body was left alone by the animals, it portended evil.[129] Letting animals determine if a human was alive or dead was used by a number of cultures. According to d'Ablaincourt, these included the peoples of the Caspian area, the Iberians, and the Hyrcanians (along the southern edge of the Caspian Sea), who kept dogs for this express purpose (sepulchral dogs).[130] (See also the following account of the Samburu people in Kenya.)

Africa (Kenya)

Among the Samburu people, the word for death (*lkiye*) means "separation" or "cutting." A dead person has had his heart "cut" (indicating that the heart is a critical indicator of life) or has been "thrown away" (into the bush). Corpses are covered with leaves and left in the bush. Death is known by its smell, since decay comes quickly in Africa. Thus we find another culture in which putrefaction is the gold standard.[131] Signs of death: dry eyes, smell, hyenas eating the body.

Even though the dead are "cut" from the living, it's possible for people to be dead and then return (resurrection/premature death). A firsthand account by a tribal woman happened sometime between 1922 and 1936:

> Wasn't she taken away to be disposed of? . . . She was laid. Isn't it that a person is usually laid? . . . She lay there. The hyenas didn't eat her. . . . Her brother, a Lkileku Imurran, got up. . . . Is it not that people are usually checked to see whether hyenas have eaten them? . . . He went, and as he went he found her just sitting like this, blinking her eyes. . . . She just kept her eyes open like

> mine now. Don't the eyes get dry when a person dies? So she was just blinking her eyes.[132]

Among these people, death is certain only after hyenas eat the body, and the people check to see if this is happening. Hyenas eat carrion, though they hunt and kill prey as well. They have a keen sense of smell, and perhaps they can smell if the person is still alive.

Melanesia

According to anthropological studies in the 1950s, peoples from the Solomon Islands and Santa Cruz delayed burials, like many other peoples. Persons of rank were grieved over for two days; persons less lofty got one day; and persons of no repute were buried immediately. All this can be interpreted as making sure that important people were not buried prematurely. For two days, women wept over the body and people came by to look at it. Then it was buried. On Banks Island and the New Hebrides, the people watching over the corpse "bite the finger of the dead or dying person to rouse him, and shout his name into his ear, in hope that the soul may hear it and return."[133] On one island, the body was placed near slowly burning fires for a full ten days and watched by the women "till nothing but skin and bones was left." This was done as "a mark of affection," but at least initially it would have functioned as a final proof that the person was dead.[134]

BACK TO THE WEST

As we entered the twentieth century, a simple test for who is dead and who is alive had not really been nailed down. Yes, a cardiopulmonary standard had gained support, but it was never that simple. In the next chapter, we will see that all of the work of the preceding five millennia was thrown out the window. Doctors in the twentieth and twenty-first century are fond of saying that cardiopulmonary failure has been the age-old standard for death. But as we have seen, that was not the case. In the late nineteenth century, many experts cited the need to set up many hurdles to jump before calling a person dead: cessation of heartbeat and breathing but also lack of reflexes, sensation to pain, circulation in the eyes, rigor mortis, livor mortis, and other symptoms. The only sure accepted sign was putrefaction, and even that could be mimicked by gangrene. Five thousand years of staring at dead and dying bodies left us with the impression that life is resilient and death may not come easily.

THREE

The Brain-Death Revolution

Harvard lowers the bar for deadness.

IN 1968, thirteen men gathered at the Harvard Medical School to virtually undo five thousand years of the study of death. Doctors, scientists, anatomists, and others who had examined death for five millennia had discovered that death is as uncertain as the mystery of life. Then, in a three-month period, the thirteen men of the Harvard committee (full name: the Ad Hoc Committee of the Harvard Medical School to Examine the Definition of Brain Death) hammered out a set of simple criteria that today allows doctors to declare a person dead in less time than it takes to get a decent eye exam. A good deal of medical language was used, but in the end the Harvard criteria switched the debate from biology to philosophy. You are dead not when your heart cannot be restarted, you can no longer breathe, or your cells die, but when you suffer a "loss of personhood."

Despite popular belief, cardiopulmonary failure was not relied upon before brain death. In the nineteenth century doctors and others also looked for rigor mortis, algor mortis (a fall in temperature), livor mortis (skin discoloration, as uncirculating blood pools), and putrefaction. As we have seen, the public was well aware that death was sometimes wrongly declared for victims of lightning, diabetic ketoacidosis, epilepsy, stroke, drowning, or even hysterical fainting. Coffins were built with elaborate escape devices, and Mark Twain reported that morgue attendants in Munich connected a wire from the allegedly dead person's finger to a bell in the morgue to alert them if the declaration of death was premature. The electrocardiograph, invented in the nineteenth century, was used to record electrical events emanating from the heart, though the EKG was not used to diagnose death. "Most patients were pronounced dead long before electrical silence of the heart occurred," states Michael A. DeVita, MD, of the University of Pittsburgh Medical Center.[1]

The first half of the twentieth century saw significant advances in technology. The electrical defibrillator was invented, as was the iron lung, which led to the modern ventilator. The EEG (electroencephalography) recording provided a view, albeit superficial, of the brain's electrical activity. Organ transplants became a reality. Two of those developments, ventilators and organ transplants, are pivotal to our story.

The need for the ventilator was highlighted at the turn of the century by the work of the American neurosurgeon Harvey Cushing, who in 1903 first described the phenomenon of "brain death." He asked himself the question: are patients with fatal brain damage dead, or are they simply on the road to death? He discovered that the answer was the latter. If he could keep their lungs breathing, the heart would follow along. Brain death was merely a serious form of coma. Cushing had to use laborious old manual artificial respiration, but even with that technique he kept one patient alive twenty-three hours after brain death. In a sense, Cushing had expanded the boundaries of life and raised the bar higher for declaring a person dead.

Keeping patients with catastrophic brain damage alive for long periods of time using manual artificial respiration was impractical, but Cushing had conducted an important experiment that was to have practical applications by midcentury. The driving force was not a scientific thirst for knowledge but a frightening disease: polio. Human beings are wired up in a strange way. Our brain stem must send signals down to the body to instruct us to breathe. The heart, on the other hand, has a brain of its own, a natural cellular pacemaker. Yes, it does take orders on occasion from the brain. Excitement generated by the presence of a predator or a sex partner can send hormones to the heart to make it beat faster. In general, though, the heart will beat quite nicely, brain or no brain.

The polio virus can paralyze the muscles needed for respiration. The patient's brain stem sends a signal to breathe, but the muscles cannot respond. The first artificial respirators developed were the very bulky iron lungs. Patients' bodies were sealed in a large cylinder with their heads sticking out one end. The iron lung worked on negative pressure, which is how the body naturally breathes. The pump reduces the pressure, the chest cavity expands, as do the lungs, and air is sucked in. When the pressure increases, the lungs are forced to exhale. But iron lungs are inconvenient for everyone involved and were slowly replaced in the 1950s by smaller ventilators, which work by blowing air into the patient's lungs via intubation through the airway. For those with brain injuries, the problem is the opposite of that of polio. The muscles of the breathing apparatus in the thorax are functioning, but the proper signals to breathe are not being sent down by the brain. By the 1960s, intensive care units (ICUs) began filling with patients who could be kept alive on ventilators.

During that era, there were also great advances in organ transplantation, both in surgical techniques and in the development of immunosuppressive drugs. There were plenty of patients who wanted organ transplants but then as now, far fewer dead people to take them from. I should rephrase that. There are always plenty of dead people, but for the purposes of transplant they are often too dead. What was needed were more people who were sort-of dead, whose hearts were still pumping and keeping those valuable, desirable organs fresh. It was against that background that the Harvard committee met in 1968.

THIRTEEN MEN, REVERSING 5,000 YEARS OF HISTORY

For five thousand years the bar for the declaration of death had been raised incrementally. Every time a telltale symptom of death was found (no breathing, no heartbeat, discoloration, etc.), an exception was later discovered. As we have seen, Cushing raised the bar higher with brain death, which meant that while the brain might be nonfunctioning, the person was still alive. In 1952, French neurologists termed brain death *coma dépassé,* or beyond coma. It was a state worse than coma but not death.

Why did the Harvard committee decide to lower the bar for determining death? What was the overwhelming scientific need? Was there in fact any scientific motivation? The answer seems obvious to the committee's critics and to anyone who has read its report. We shall get into the committee's not-so-hidden agenda later. At the moment, let's examine its conclusions, as put forth in its paper in the August 5, 1968, edition of *The Journal of the American Medical Association* (*JAMA*) entitled "A Definition of Irreversible Coma."[2] Irreversible coma is what the committee preferred calling brain death, which as we'll come to see, is an unfortunate name for those who support it as legitimate death.

The first sentence is straightforward: "Our primary purpose is to define irreversible coma as a criterion for death." Then the committee stated its reasons for a new definition of death:

> (1) Improvements in resuscitative and supportive measures have led to increased efforts to save those who are desperately injured. Sometimes these efforts have only partial success so that the result is an individual whose heart continues to beat but whose brain is irreversibly damaged. The burden is great on patients who suffer permanent loss of intellect, on their families, on the hospitals, and on those in need of hospital beds already occupied by these comatose patients.

In other words, the committee members felt bad for the families of patients in severe coma, and they proposed to make the families feel better by killing their loved ones. In addition, it would be convenient to clear out some of the beds in the ICU. The committee went on to state a second reason:

> (2) Obsolete criteria for the definition of death can lead to controversy in obtaining organs for transplantation.

This statement is even more baffling. What controversy is the committee referring to? Prior to 1968 only kidneys were taken from living donors. Otherwise you could not harvest the organs of a living person, one who was breathing and whose heart was beating. The so-called Harvard criteria were attempting—and eventually succeeded—to change that. Taking organs from pink, breathing people did not assuage any controversy; it created one.

It is normally not prudent to begin an article, especially a scientific paper that will redefine who is alive and who is dead, with two baldly incredulous statements. We will return to such problems later. For now, let us get on with the Harvard paper. The committee was reasonably succinct, its article taking up only three and a half pages in *JAMA*. The authors quickly moved to set down the three, or possibly four, criteria that indicated that a patient had a "*permanently* [emphasis the committee's] nonfunctioning brain."

- **The patient must be *unreceptive* and *unresponsive*.** Echoing research from eighteenth-century Europe, the committee said, "Even the most intensely painful stimuli evoke no vocal or other response, not even a groan, withdrawal of a limb, or quickening of respiration."
- **No movements or breathing.** The committee instructed doctors to watch patients for at least one hour to make sure they made no spontaneous muscular movements or exhibited spontaneous respiration. To test the latter, the physicians are to turn off the respirator for three minutes to see if the patient attempts to breathe on his own. This is called the "apnea test."
- **No reflexes.** Doctors simply look for reflexes, shine a light into the eyes to make sure the pupils are dilated. Muscles are tested. Ice water is poured into the ears. We will go into detail later.
- **Flat EEG.** The committee also recommended that doctors use electroencephalography to make sure the

> patient has flat brain waves. The committee stressed the importance of this confirmatory test.

The committee said all of the above tests had to be repeated at least twenty-four hours later with no change. It went on to state that the decision to declare a patient dead was the domain of the physician in charge and that the patient's family should have no say in the matter. The committee stated this in a positive way, as if doing the family a favor: "It is unsound and undesirable to force the family to make the decision." The article made an interesting statement: "Death is to be declared and *then* the respirator turned off." (As we shall see, this rarely happens. After death is declared, if the patient is a potential organ donor, the respirator is immediately reconnected.)

The committee added two caveats at the end of the article. It pointed out that hypothermia and drug intoxication can mimic brain death. In the many years since the 1968 article, the list of mimicking conditions has grown longer. As a writer who has covered science for four decades, this doesn't sound quite like science. Physical laws don't usually have exceptions. When Isaac Newton fashioned his formula $F = ma$, or Force equals mass times acceleration, he didn't footnote a long list of exceptions (formula doesn't work for pineapples, Buicks, Norwegians . . .).

Granted, the Harvard committee was attempting to formulate not a physical law but criteria. Still, the members were deciding who should be allowed a ventilator and who should not, and one would like to see a bit more rigor. Another disappointment is that the Harvard criteria are based on zero patients. No experiments were conducted with either humans or animals. It was all theory, the thoughts and desires of the thirteen men on the committee.

There were also no citations to data. In all fairness, there *was* one citation, to a speech by Pope Pius XII, who argued that the only people who should be given the power to declare death are doctors. The committee, eleven of whose thirteen members were doctors, agreed. Pope Pius must have forgotten that he himself, as had been the custom for centuries, would be declared dead by the camerlengo, a special nonmedical cardinal whose technique is to hit the pope in the head with a hammer. There have been many criticisms of the committee's attempt—successful as it was—to spin-doctor coma, no matter how irreversible, into death. As many have asked: how does a dead person sustain a coma?

No matter. The Harvard ad hoc committee article remains the ur-text of death determination. The Harvard criteria had no legal standing but incredible clout. They soon became the standard for declaring people dead in several states, and in 1981 the Uniform Determination of Death Act (UDDA) was sanctioned by the National Conference of Commissioners on Uniform State Laws. The UDDA is based on the Harvard ad hoc committee's report. That a three-page article, with no data, no patients, and only one citation—to a pope—should be codified by all fifty states within thirteen years seems miraculous. *Somebody* in the medical community wanted it to succeed. I'm guessing, too, that the name "Harvard" at the top of the article might have helped, as opposed to, say, "University of Southern North Dakota A&I."[3]

One of the less celebrated studies is commonly known as the Minnesota criteria, published three years after the Harvard criteria. Its real name is "Brain Death: A Clinical and Pathological Study."[4] Perhaps because its formal name bears little resemblance to its common label, the study appears to be rarely read though often cited, usually incorrectly. Its objective was to test—on real patients—the four general criteria recommended by the Harvard committee. The Minnesota Health Sciences Center studied twenty-five patients, twenty of whom progressed to brain death due to head injury.

To sixteen of these patients, only the first three criteria were applied: the clinical and apnea tests. The EEG was skipped. EEGs were performed on nine patients, and if you believe most of the literature on the subject, two of those patients had positive EEGs. This is not good. If a person's brain is dead, it shouldn't be sending out brain waves. Even two out of nine patients with positive EEGs are frightening. There shouldn't be any. But it was worse than that. If you ignore the many citations to the Minnesota criteria and read the actual study, you'll find that the Minnesota team found electrical brain signals in five of the nine patients—more than half. In three of those patients, the brain waves were detected either twenty-four hours or two days before they were declared dead, so they disregarded them. In those three cases, however, a second EEG twenty-four hours later, as recommended by the Harvard criteria, was not done. The team also had to give one of the other four patients who were considered flatliners seven consecutive EEGs before it got him to stop transmitting brain waves.

Whether you consider that five brain-dead patients had EEGs or only two, two is two too many. If you are finding any dead people with brain waves, what's waving? The Minnesota study falsified[5] the Harvard criteria. Even if one assumes that seven patients flatlined on the EEG or would have flatlined had they been tested, that does not compensate for the two patients who still had brain activity at the time death was declared. Those two patients showed that the criteria were not working, and when you are deciding who is dead and who is alive, seven right out of nine is not an acceptable score.

In any other scientific arena the Harvard criteria would have been dismissed. Instead, the Minnesota study inexplicably concluded, "Except in unusual circumstances, EEG need not be a required procedure for the neurosurgeon in the determination of brain death."[6] In other words, the EEG put the lie to the Harvard method, and the solution was simple: get rid of it. A common defense of the Minnesota solution is that the researchers later did autopsies on all twenty-five brains. But the paper details only damage done to the brain stem, which is not what EEGs test. The EEG's sole purpose is to look for activity in the cortex, the higher centers of the brain. The authors conclude that insistence on EEG examination "would be unduly restrictive on the practice of medicine."[7] The Minnesota researchers also discovered that brain-dead people move a bit, another trait forbidden by the Harvard criteria. More about that in the next chapter.

A constant thorn in the side of the pro-brain-death forces is the term itself. "Brain death" doesn't sound dead enough. The term implies that only part of the person, the brain, is dead, but the human as a whole is still alive. One doctor cites a newspaper headline: THREE AGENTS SHOT IN DRUG BUY; 1 KILLED, ANOTHER BRAIN DEAD.[8] Or in jokes we often refer to a stupid person as brain dead. We don't mean he is dead, just that he is so dim it *seems* as if his brain is dead. In effect, the Harvard committee and its supporters were making a philosophical/religious statement. A man *is* his brain. When that is gone, he is gone. Or it could be stated: brain = mind = person. Nothing like this was written in the Harvard report, but it is certainly implied.

The other problem with the term "brain death" is that it created two kinds of death, in that cardiopulmonary death was still on the books. This annoyed the brain-death forces, some of whom suggested changing "brain death" to simply "death." It was, after all, they felt, the gold standard of death. There can be only one kind of death, they said. But Alan Shewmon, a UCLA pediatric neurologist, says, "The term itself is kind of ambiguous from the start, because in one sense it could simply mean death of the brain as an organ, just like you can have necrosis or death of a finger if you cut off the blood supply to it. So the brain itself can die due to lack of blood supply."[9] But, asks Shewmon, does it necessarily mean death of the whole person? In any case, there were now two sets of criteria. Unfortunately for brain-death advocates, most people preferred the cardiopulmonary criteria. Lots of people could detect failure of the heart, circulation, and breathing, and it was difficult to accept that a brain-dead patient, who was still breathing and had an active heart, was no longer alive.

THE APOSTLE PAUL OF BRAIN DEATH

One of the champions of brain death was the late Julius Korein, a neurologist at New York Medical Center. (It is probably not a coincidence that proponents of brain death include many neurologists, who would be prone to believe that the brain and the central nervous system would determine "personhood.") Korein is perhaps best known for testifying in the Karen Ann Quinlan trial in favor of her parents, who wished to disconnect their daughter, who was in a persistent vegetative state (PVS), from her ventilator. In April 1975, the twenty-one-year-old Ms. Quinlan came home from a party, where she had been drinking and taking drugs, including Valium, and collapsed and stopped breathing for fifteen minutes or more. Taken to the hospital, she lapsed into PVS. Her father sued to take her off the ventilator, so she could "die with dignity," and Korein and other neurologists testified on his behalf. After several legal skirmishes, Mr. Quinlan won before the New Jersey Supreme Court.

Korein's opinion, and that of his colleague Fred Plum, was that once the ventilator was removed, Karen Ann Quinlan would move directly to brain death. Instead she began to breathe on her own and, instead of "dying with dignity," lived almost ten more years, dying in 1985 of pneumonia. This diagnostic failure evidently did not dampen Korein's confidence in himself. He went on to edit the bible in the field, *Brain Death: Interrelated Medical and Social Issues,* a 454-page collection of articles and essays by experts in the field published as volume 315 of the *Annals of the New York Academy of Sciences.*

Many of the people I interviewed referred me to Korein's book to answer questions I had about the scientific legitimacy of brain death. It reminded me of sitting in catechism class, being lectured about the meaning of the Bible by a minister who never seemed to have a Bible at hand or appeared to have read it. I find it hard to believe that more than a few doctors have read beyond Korein's optimistic preface, if that.

Like others, Korein appears out of sorts over there being "two different kinds of death." Like others, he says brain death "does not replace the classical definition of death but only adds another set of criteria." He then does some fancy verbal footwork in an attempt to show that both cardiopulmonary criteria and the Harvard criteria are "predicated on an implicit *concept* of brain death." His argument is that when there is "irreversible total circulatory and respiratory failure, death of the brain inevitably follows."[10] It's just that with brain failure, the brain goes first.

Korein's logic might also be applied to the liver, kidneys, or any other organs. His assumption is that the brain equals the person. Korein also specifies how remote technical brain death is from death of the heart. He admits that literal death of the whole brain is not required for a declaration of brain death. He states: "It is not required that every neuron in the brain be destroyed." Brain death only means that "all neuronal structures in the brain *will* die" (emphasis mine). He writes that "cardiac arrest and death of the adult human being is inevitable within one week." Thus, he says the two definitions of death (cardiopulmonary and brain) "conceptually define death in the same manner, that is, death of the brain."[11] Except that brain death predicts only that the person *will* die, with a waiting period of up to a week. Which seems to undercut brain death as death per se.

503 AUTOPSIES

Most noteworthy in editor Korein's compilation is a study from the National Institute of Neurological Disorders and Stroke (NINDS). Between 1970 and 1972, soon after the Harvard article was published, the NINDS researchers followed 503 comatose patients from the time of onset of apnea. The study found that patients who fit the Harvard criteria *will* probably die but are not necessarily dead at the time they meet the criteria.

Detailed autopsies did not always show signs of autolysis (self-destruction) of the brain or other features of death. Only 4 percent of subjects met the Harvard criteria before spontaneous cardiac arrest or termination of life support. Which means that in 96 percent of cases, brain death is irrelevant as it comes *after* cardiopulmonary death. As in the Minnesota study, there was a fraction of patients—in this case 17 out of 503—who met all of the Harvard criteria with the exception of the EEG. Those 17 showed biological activity in the EEG at the time of death.[12]

Recall that the Harvard committee included the EEG in its original criteria, but the Minnesota criteria threw it out. The NINDS study, which involved nine institutions all putting in "three years of the most gruesome work," according to one doctor, employed the EEG on every patient. *Time* magazine greeted the results with characteristic optimism, writing that the NINDS had streamlined the organ procurement process. No longer did the brain-death test need to take twenty-four hours. The NINDS program ascertained patients' deaths in only thirty minutes.[13] That turns out to be a misreading of the data. The thirty minutes refers only to the EEG. If the patient recorded half an hour of flat brainwaves, the NINDS said he was not going to survive. The patient was not necessarily dead, but he was not going to recover. The Harvard committee originally required only ten minutes of flat EEG, as opposed to thirty, to be repeated twenty-four hours later. All of this is academic because the EEG was dropped from all U.S. criteria in the 1970s.

In defense of *Time* magazine, journalists were very pro organ transplant in the 1970s and still are. I myself ran a favorable article telling people how to become organ donors when I was an editor at *Good Housekeeping* in the mid-1970s. Transplantation was sexy, space-age medical technology, a dream, and it was happening in our own time. Arthur C. Clarke said, "Any sufficiently advanced technology is indistinguishable from magic." This seemed like magic. As a science writer, I wanted it to happen and be successful, just as I wanted the Apollo program to succeed. We thought in those days about the wondrous thing that was happening to the recipients. We read about them. We knew their names. We didn't think much about the donors. They were dead, after all, and there was nothing to worry about. We figured we could trust doctors to figure out when someone was dead. "Transplant technology motivated people," says DeVita. "Envelopes were pushed."

In fact, the NINDS study was not beneficial to the brain-death concept or organ transplanation. Gaetano F. Molinari, in the Department of Neurology at George Washington University

Medical Center in Washington, D.C., writing in Korein's bible, was clearly shocked by the seventeen patients who had brain-wave activity despite having been judged brain dead by all the other criteria. "These criteria," wrote Molinari, "permit a margin of error that is unacceptable." The criteria used by the NINDS researchers also appear to be far more rigorous than those recommended by the Harvard group. The clinical exams took an average of seven hours to perform. The weight those doctors gave to the EEG is worrisome, given the fact the EEGs were, and still are, no longer a requirement.

Molinari found the Harvard criteria wanting: "Since pathologic findings [in the NINDS study] did not always confirm brain death, even in patients meeting the more stringent Harvard criteria, the end-point or proof of validity of these criteria remains ill-defined. Prediction of a fatal outcome is not a valid criterion for accuracy of standards designed to determine that death has already occurred."[14] In other words, the Harvard criteria predict that the patient will die. They do not tell whether the patient is dead.

Molinari suggests that more technology should be used at the bedside, especially tests that assess blood circulation in the brain's cortex, which as we will see, is rarely tested. But, he goes on, both doctors and scholars have resisted the idea of using more technology, and he noted that even the EEG is not available to all physicians called upon to pronounce death. Korein himself recommended, in a 1980 letter to the *British Medical Journal*, that brain-death criteria require not only an EEG but also studies of cerebral blood flow.

None of these would come to be. Both Molinari and Korein state clearly that the brain-death standard—and the need to declare it quickly—are a consequence of the need for organs to transplant. The NINDS team also found that responsible doctors had missed the fact that some of their patients had taken drugs that made their bodies mimic brain death.[15] Also hiding in the report is the fact that the NINDS scientists observed reflexes, spontaneous and induced movements, and muscle tone in some patients. The Harvard committee had said that dead people don't move. All this in a book that supposedly endorses brain death.

The EEG is not exactly high tech. It looks for electrical signals on the scalp. There could be a storm of brain waves in the interior of the brain's cortex that EEGs do not measure. As the University of Pittsburgh's DeVita said, "It's like looking at the moon with binoculars."[16] Shewmon points out that "some patterns for 'brain dead' look a lot like sleep. But physicians call it brain death."[17] In the early decades of brain death, there was talk throughout the literature of adding more sophisticated tests: auditory brain stem response (ABR), four-vessel angiography, radionuclide methods, in which an isotope is injected into the bloodstream, images of blood flow to the brain, and others. There is no point in discussing them further, as such tests, even the EEG, are rarely used. It was assumed that blood flow studies would become more prominent within a decade. That has yet to happen.[18]

FLASH-SPLASH-GASP: REAL-WORLD BRAIN DEATH

Cooper University Hospital in Camden is home to one of New Jersey's three Level 1 trauma centers. Cooper serves the critically injured from the nine counties of south Jersey. There are lots of dangerous back roads, and Atlantic City beckons from the east, drawing happy, expectant drivers in one direction and spitting out broke, depressed motorists in the other. The traumatically injured arrive via ambulance and helicopter. In fact, 40 percent are medevaced in. With 2,500 unhappy customers a year, Cooper is one of the great brain-death capitals of the world.

Steven Ross, MD, is in charge of the trauma unit.[19] The majority of trauma deaths result from bleeding out. Ten to 15 percent of patients who show up at his trauma center are dead on arrival (DOA) of sepsis or bleeding. The brain dead are in the minority. The day before I talked to him, he had treated a man with a gunshot wound to the head. He had no sign of brain function. Ross was in luck. The patient's family is in favor of organ donation.

The picture the public has of space-age technology being hooked up to patients to ascertain their deaths is highly unfounded. Ross explains that testing for death is pretty simple. Basically, the test requires a Q-tip, a flashlight, some ice water, and a rubber hammer. It is primarily a clinical test, usually performed in a hospital because the patient is on a respirator, but most of it could be done in a country doctor's office. As Ross explains, the Harvard criteria—minus the EEG—consists of the following:

- **Corneal test.** He touches a cotton swab to the patient's eye, looking for a blink, a sign of life. (In most cases the doctor also uses a flashlight to examine the eyes.)
- **Gag reflex.** He pulls out the tube in the throat, looking for a gag.
- **Suction in the throat.** To see if the patient coughs.
- **Cold water poured into the ears.** If the patient is alive, his or her eyes will shiver and move back and forth.
- **Doll's eyes.** He rocks the patient's head back and forth. If the patient is dead, his eyes will move with his head, like eyes painted on a doll.

- **Knee reflex.** This is not a required test, but Ross does it anyway. Even if the patient's knee jerks, he can still be declared dead.

If the patient "passes" all of these tests—which is to say he passes for dead—the apnea test is performed. The ventilator is disconnected to see if the patient can breathe on his own. If he gasps for air, he is still alive. Basically, the brain death exam is flash-splash-gasp: a flashlight in the eyes, ice water in the ears, and then an attempt to gasp for air. In New Jersey, says Ross, all of these tests are repeated six hours later. Only then—legally—can the patient be declared dead. Two different doctors must be used, and one must be a neurologist or neurosurgeon.

While a patient cannot be declared brain dead until the second set of tests, Ross says he or his staff inform the organ procurement organization (OPO), or organ bank after what he calls "the first brain death." The implication here is that the doctor thinks of his patient as dead many hours before he is legally so. And the OPO gets an early shot at obtaining the body.

Ross will not use EEG because, as the Minnesota criteria decided, "Brain dead people look alive" when EEG is used. Using a word popular when talking about EEG, he says, "EEG is prone to *artifact*." Also, the Harvard criteria mandate that the EEG be repeated twenty-four hours later, and that would delay the confirmation of brain death. I said that a Level 1 trauma center, such as that at Cooper University Hospital, must have a healthy number of brain-imaging devices. He said sure, it has stop-flow scans, nuclear medicine, radioactive imaging, and other technology, but they rarely use such equipment. He estimated that only two such tests per year are performed, and only when requested by the family.

When the patient passes—or fails, depending on how you

see it—the second set of tests, including the apnea test, and he is declared brain dead, one of two things can happen. If the patient does not become an organ donor, Ross takes out the ventilator and waits for the heart to stop. In other words, if you're not an organ donor, you are allowed the luxury of plain-vanilla death, cardiopulmonary death. In fact, many hospitals use the time the heart stops, not the brain-death declaration, as the time of death. But if the family is giving up the body to the OPO, it is legal for him to "keep the body alive." That is, once the patient flunks the final hurdle, the apnea test—he fails to gasp for breath—the ventilator does not remain disconnected. Rather, it is reconnected so the dead body can, in Ross's paradoxical words, stay "alive." As Ross puts it, "There is a big change in treatment when the patient dies." In most cases, he will now get the best medical care of his life, even though that life is, from a legal standpoint, technically over. (The language gets difficult here.) Ross says that keeping the body "alive"—creating a "beating-heart cadaver," in the parlance—is not allowed if permission hasn't been granted. But if it has been granted, there is no legal time limit on how long one can keep the body alive, though it is difficult to do so.

Ross likes to assess the attitudes of the patient's family. Are they open to donation? New Jersey has a "conscience clause"; people can object to brain death on the basis of their religion. Ross said he's had only one such case. He forgets the family's exact religion, but it was an offshoot of the Pentacostals. Ross's policy is "Don't ask, don't know."

I asked Ross if doctors are generally sympathetic to organ banks. "Very much so," he said. I found out later that he is a member of the New Jersey Organ & Tissue Sharing Network Board of Trustees. The Harvard Ad Hoc Committee had stated that transplant doctors should not be involved in proclaiming brain death, but the relationship between hospital staff and OPOs has become intertwined, as we will see in greater detail in chapter 4.

Finally, Ross defended brain death on philosophical grounds: "This person's soul is no longer here. What made this person a person is no longer there."

NO LIABILITY

The public is generally unaware of how quickly a patient can be pronounced brain dead. Even the lawyer Alan Dershowitz told me that brain death requires an EEG, which it does not. I asked a number of laymen at random what "brain death" meant to them, and they all mentioned a flat EEG or "absence of brain waves" or something on that order. I attended a panel discussion at Hampshire College in Amherst, Massachusetts, called "Matters of Life and Death." On the panel was a Hampshire professor of psychology who told the audience of mostly students that the EEG was required to determine brain death and that his mother, who had died in a hospital in a rural Wisconsin town, had had a flat EEG. I found this amazing, that a backwater hospital performed EEGs when major trauma centers did not.

I talked to him afterward amid a throng of fawning students and asked if he had actually seen the EEG performed on his mother. He said no, but he was sure the test had been done. I asked why, and he said because the Harvard criteria require it. I asked him if he realized that the EEG had been eliminated as unnecessary back in 1971. That startled him, and instead of answering me, he literally turned his back on me and his students and ran out of the auditorium. I had a son in college at the time, and I felt sorry for the parents who had offspring being taught by such uninformed professors and were paying $50,000 a year for such misinformation.

The Uniform Determination of Death Act (UDDA), approved in 1981, is fairly short, comprising fewer than two typewritten pages. The essence is that "the entire brain must cease to function, irreversibly." It goes on to state, "This act is silent on acceptable diagnostic tests and medical procedures. . . . The medical profession remains free to formulate acceptable medical practices and to utilize new biomedical knowledge, diagnostic tests and equipment." The law's drafters wanted a general rule that could be applied to small hospitals in, say, Ely, Minnesota, as well as to Massachusetts General Hospital. What has happened is that the lowest possible standard tends to reign.

The UDDA ends by stating that doctors who declare death cannot be prosecuted criminally or be "liable for damages in any civil action" as long as they act in "good faith."

FERNANDA'S LAST FEW MINUTES

We wanted to see brain death for ourselves. "We" was Judith Hooper, my cowriter on the original magazine article that sparked the research for this book, and I. A doctor at the University of Pittsburgh, possibly the leading U.S. center in organ transplantation, set us up with what he called the best brain-death team in Massachusetts, led by Daniel Teres, MD (no relation), an intensivist and chief of critical care at Baystate Medical Center for twenty-three years. Hooper and I split the week into various shifts, and Baystate, in Springfield, was to call one of us whenever it had a potential brain death. We were put into the same ghoulish position as the organ banks, waiting breathlessly and hopefully for another human being to be struck by neurological catastrophe.

On my shift I got a call about a young woman, a Smith College student, who had been mowed down in a crosswalk near the college's main entrance in Northampton. The town is usually very pedestrian friendly, with a $100 fine for anyone driving through an occupied crosswalk. It's a policy the town believes stimulates business, as shoppers can easily get across the street to other stores. Also, live shoppers tend to purchase more than dead ones. The president of Smith College told me that it's not even necessary to use a crosswalk: cars will stop for any pedestrian who steps out into the street. This student was an exception. Before the brain-death team could be assembled and I could drive to the hospital, the student's heart stopped, and she suffered conventional death. I was disappointed, but probably not as much as the organ bank.

Another call from Baystate, which occurred on Hooper's shift, was more fruitful. This was about Fernanda, whom we met at the beginning of the book. At fifty-three years of age, she was found lying on the floor unconscious. When she got to the emergency room she was "unresponsive," her pupils dilated and nearly "fixed." A CT scan showed that a clot had floated up from her heart to lodge in her medial cerebral artery (MCA). She had suffered a stroke of the most severe kind. Her ischemic brain had started to swell. Doctors felt it was only a matter of time until her brain would herniate and then become squeezed against the skull until all of her mental life was irrevocably destroyed.

Fernanda was single, never married, and now under the care of Teres's unit, this particular team headed by Dr. Thomas Higgins. Hooper met Higgins outside the ICU and audiotaped the entire procedure. Other witnesses included an entourage of hospital "orientees," as they were called. Meanwhile, Fernanda's family had been sequestered in a private room. The following is Hooper's transcript, only slightly edited. I've added emphasis in the form of underlining in some places to denote key parts of the brain-death exam.

HIGGINS [briefing Hooper]: Lady came in a couple of nights ago with ischemic stroke in her left middle cerebral artery distribution—devastating if you're right-handed—controls right side of body. Glasgow coma scale of six at best. Scale goes from three to fifteen, where fifteen is normal and three is very abnormal. So six is not good. Because of extent of stroke, the size of it, it was felt that blood thinners were not indicated here. The prognosis was poor.

HOOPER: The damage was so extensive?

HIGGINS: Yes. With a smaller ischemic stroke sometimes you can anticoagulate and open up blood flow to the area. It's not that you really save the brain tissue at the center of the stroke, but you can sometimes save enough of the surrounding brain tissue from progressing to a full stroke. There is an ischemic penumbra, or an area surrounding the actual stroke.

And with a stroke this large you can have progressive swelling of the brain . . . that swelling will eventually cause herniation of the brain. In other words, there is not enough room for the brain to be contained within the cerebral vault, so it starts to push downward and compresses a large part of brain in a small space.

[Sound effect: steady chime, *ding-ding-ding*, like an old-fashioned elevator.]

Our initial exam here showed she had brain-stem reflexes. They were not normal. She had a very weak gag and had lost the right corneal reflex but not the left. And she was breathing on her own. She was breathing over the vent rate. From this we know she has a severe neurologic injury but she's not brain dead.

So I met with the family today, told them there were three possibilities. One, swelling in brain continues, she becomes brain dead; two, stays where she is or makes a small improvement, but we don't know whether she will ever improve or the extent of that improvement; or, three, based on her previously expressed wishes and the family's knowledge of her, we withdraw support in what we call "terminal weaning." We take her off the ventilator, and whatever happens happens. If she breathes on her own, we provide her with comfort measures. If she doesn't breathe, we don't do anything to change that. There's a small possibility that she could breathe on her own and continue in this persistent vegetative state for months, typically days—typically it's hours to days, but I've seen patients go for six months to a year. We've already contacted the NorthEast Organ Procurement Organization, but they haven't evaluated her yet. I have not talked to the family yet.

DENNIS O. [a nurse, arrives]: She's finished. Nothing there. Nothing on her neuro exam.

HIGGINS: Okay, so this is a change. When did she have the Cushing's reflex?

DENNIS: Four.

HIGGINS: So what's probably happened here is that the swelling has increased. Dr. Harvey Cushing, a neurosurgeon around the turn of the [twentieth] century, described a reflex for increased intracranial pressure, and it has since become known as Cushing's reflex. It consists of hypertension and bradycardia. So the blood pressure goes up and the heart rate goes down. It's a sign of impending neurological catastrophe. She did that at four o'clock while we were on rounds. Dennis, who is a very good nurse, has done a quick informal exam, and I'll go in and confirm his findings. So it sounds like she may be brain dead.

[We all move into Fernanda's room.]

HIGGINS: First thing on brain-death evaluation is the prerequisites. We have to ensure that the patient doesn't have any drugs on board—sedative agents. Second thing, blood pressure has to be adequate. You need a systolic of at least 90. Hers is 124. The third thing is temperature has to be at least 95 degrees. There are instances of drowning victims who have been in cold water and have appeared to be dead, and when we warm them up they revive. [He checks Fernanda's temperature.] Temperature, 97.5. The next order of business is there are four reflexes we check, starting wtih corneas. We'll skip the Q-tip for a minute. We use that for the corneas. What I need is the piece of the otoscope. We want to make sure that there is no wax in the canal obstructing the tympanic membrane and no perforations in the tympanic membrane. In her it's unlikely, but with trauma victims there can be a perforation and you don't want to be pouring ice water over the ear drum.

[Dennis brings in part of the otoscope; Higgins looks in her ears.]

Her tympanic membrane is completely normal, no perforations, no wax in the ear. Same thing on this side, looks good. Okay, you know those penlights that everyone carries around? Worthless. [Picks up a full-size flashlight.] Okay, I'm going to check the pupillary reflex. Bring the light over, and I'm going to look to see if the pupils change size, and they don't, so her pupils are fixed and dilated. Her pupils are about four to four and a half millimeters in size now, so they're dilated. Someone who's surprised or in a dark room might have pupils like that, but with bright light like this they should be down around three.

Second thing is corneal reflex. I'm going to use a sterile swab here. The only reason for that is she may be an organ donor for a corneal transplant and we don't want any infection. So bring it to the side . . . and look to see if there's any blink reflex . . . and there's none. Other side as well. I touch the cornea: she should blink, no response.

Third thing is to look for doll's eyes. Normally there's a little bit of a delay between moving your head and when the eyes track; the eyes will tend to remain where they were and then slowly drift over; someone who is brain-dead doesn't have that response. The eyes are not going to move either way. Like a doll with painted eyes. I'm going to put her head to the midline, over to this side, and back again.

The next thing is suction. If you put this [catheter] down the back of the throat, anyone would gag quite a bit. That's another brain-stem reflex, one of the last things you lose. So I'm back here in her pharynx and she's not responding to that at all. Also—this is optimal—but if I put the suction catheter down to the point where it's going to be in her trachea where it is now, she should react to that, so she's got no gag, no cough reflex.

The final test of brain-stem reflexes is . . . ice-cold water. The vestibular system is at equilibrium with body temperature. You put ice-cold water in there, it's going to change that equilibrium. It's going to cause the fluid in the semicircular canals to move, and that's going to be interpreted as spinning. The normal response will be deviation of the eyes, nystagmus. The eyes won't move if there are abnormal brain-stem reflexes. This is a soft catheter. Dennis is going to hold the eyes open so I can see. I have previously looked in the ear canals to make sure that there was no wax, no perforation. I'm going to put in ten cubic centimeters of cold water [he uses a syringe], and I'm going to look to see if there's any response. There's not.

Some of you might be observing right now that there's a little pulsation, looks like she's swallowing. She's not. That's just her carotid. The carotid pulse is being transmitted here. So no motion on that side. And this response is immediate. If you did this on an awake person, he'd get sick to his stomach and vomit. . . . No response. So I've convinced myself she has no brain-stem reflexes.

There are a few other tests we can do, and I know these are going to be negative because we've done them earlier. You can basically rub on the chest and put some pressure on her toe. Someone who is awake . . . there should be some response to that—they'd withdraw. No response. On the fingers or on the thumb. No response. That's not part of brain-death evaluation but a quick test.

Final thing that we do to clinch the diagnosis clinically is an apnea test. Most of us will breathe if our pCO_2 [partial pressure of carbon dioxide in the blood] rises three or four points above what baseline value is. We're going to take her off the ventilator after preoxygenating her carotid. We are putting her on some oxygen so she doesn't desaturate. Two reasons to do this. One, it's good patient care. Two, if she's a potential organ donor, we don't want to compromise the organs.

It's important to note that she can't hear or understand anything we're saying now. Normally I would not talk in front of patients like this.

DENNIS: Whatever it was that made her *her* isn't there anymore. Okay, I'm going to ask respiratory therapy to come in here. I'll uncover her, take her johnny off. I'll bare her chest.

HIGGINS: Okay, Julie [respiratory therapist] is going to draw a baseline blood gas right now, and the reason is that if you've accidentally overventilated her and her starting pCO_2 is below thirty, you're going to not get a good result to test. We want to see a baseline between thirty-five and forty-five.

[Note: The reason that the carbon dioxide level is so important is that the hospital team wants to see if the patient will gasp when the ventilator is disconnected, evidence that she is still alive. Humans don't gasp for breath because they need oxygen—though they do. They gasp to expel carbon dioxide. If the pCO_2 is too low, the patient will have no impetus to breathe, even if she can.]

DENNIS: And watch her vital signs. They can get very unstable.

HIGGINS [pointing to oxygen saturation reading on monitor, in electric purple]: We will be watching this value. Anything above eighty-five is okay. If it drops below that we're going to abort the test, put her back on the ventilator. [Conventionally, to oxygenate the patient, the ventilator is put in a special mode or an oxygen cannula is placed down her endotracheal tube.]

DENNIS [regarding organ people]: I'll tell them what's happened, and then maybe they'll come down. I already mentioned it to the nephew, who is the spokesperson and the person giving consent. We already had scheduled a family meeting for tomorrow.

HIGGINS: Technically, you only need one apnea test. But everything else I did has to be repeated. I can't do the second one. I'll call one of the trauma guys and see if he can do it.

HOOPER: Is it rare to fail the apnea test and pass the second clinical test later?

DENNIS [laughs]: It has happened.

HOOPER: Does that make you doubt the apnea test?

DENNIS: No. It usually is a dealie. It just means we'll do everything again tomorrow. That person who had the questionable apnea test died later that day.

STUDENT: So if she's a donor, she would be on cardiac meds?

DENNIS: She's on Levophed already.

STUDENT: So you would have to continue with that?

DENNIS: It's a vasoconstrictor [to maintain blood pressure] for the kidneys; they need a constant blood flow through there. If her blood pressure drops, she has no blood flow through the kidneys. Her kidneys would be ruined so you have to maintain the viability of the organs.

[Someone asks about testing the patient for her suitability as a donor.]

HIGGINS: Yeah, you'll bring organ donors down for cardiac caths [catherization]. They'll do lymph nodes for cardiac typing.

[At that point the ventilator is turned off.]

RESPIRATORY TECHNICIAN: Vent's off. Got it?

HIGGINS: Okay.

[Silence.]

HIGGINS: I'm watching the chest and actually I probably will palpate . . .

[Silence except for ambient ICU noise, phones, and electronic chimes.]

HIGGINS: So she didn't make any respiratory motions at all.

[Test ends, ventilator back on. Respiratory therapists draw sample of arterial blood to test.]

DENNIS: NEOPO [New England OPO, or organ procurement organization; they pronounce the acronym nee-OH-poe] called. They're going to call back later.

HIGGINS: Funny, she was more stable during the apnea test.

Thus ended the first brain-death test. Technically, Fernanda is not yet dead. A second set of tests must be conducted by a second doctor, but this is not like the SATs. You're allowed to get a retest, but the results rarely get better. (This is possibly a self-fulfilling prophecy, but more about that later, in chapter 8.[20]) Two hours later[21] Fernanda would fail the second test also. That would count as the time of death, though for practical purposes she was dead when Higgins said, "So she didn't make any respiratory motions at all."

Julie, the respiratory technician, drew blood from Fernanda to confirm that the pCO_2 was high enough. Fernanda's baseline pCO_2 was 37 millimeters of mercury. The Baystate death team had to wait during the apnea test until the carbon dioxide in her system rose to 47 millimeters, or 10 millimeters over the baseline before disconnecting the ventilator. Otherwise her body might not have an impetus to breathe and the apnea test would not be valid and would have to be repeated. Julie announces that Fernanda hit 50.1 millimeters of mercury. The apnea test was thus valid. (Note: Other jurisdictions insist on a higher standard. The University of Pittsburgh Medical Center, for example, a pioneer in organ transplants, requires a pCO_2 of 60 millimeters of mercury.)[22]

Higgins said that Fernanda had probably had a brief headache and then fell into unconsciousness. Not a bad way to go, he said, but hard on the family, as they didn't have a chance to say good-bye. There was a bit of camerlengo in Higgins's technique. Before the above tests he had checked for unresponsiveness. "You just shout the patient's name, and you rub his chest—a sternal rub like this is very uncomfortable. People who are heavily sedated will still respond."

Hooper's lasting impression was how good Fernanda looked. Having failed the apnea test and back on the respirator, her face remained blank, unperturbed. Hooper said, "It is hard to think of her either as dead or alive, and death becomes somehow a nonevent." She noted that there is no privacy in brain death. During the apnea test the covers were pulled back, and Fernanda's hospital johnnie was pulled down to the waist, exposing her breasts while Higgins diligently placed his hands on her chest and stared at her chest and throat in search of any quivers of life. On the other hand, though the brain-death criteria may be flawed, I believe the transcript above confirms the diligence, care, and thoughtfulness of Higgins and his team.

Higgins told Fernanda's nephew what had happened. The nephew was a nice-looking, twenty-something guy in a sports shirt, his dark eyes filled with sorrow. Higgins asked him if the family had made a decision about organ donation. The nephew said that even though he is an organ donor himself—believes in organ donation—the family's feeling is that they don't know his aunt's feelings on the matter and so have to say no. Higgins later said that the rate of organ donation is much lower whenever there are language and cultural barriers. He also said that NEOPO might send someone later to attempt to change the family's mind.

After declaring Fernanda dead, Higgins said, "Brain death is pretty clear. No different from saying someone's heart has stopped and you can't resuscitate him. That to me doesn't represent any ambiguity. For most physicians it's pretty cut and dried."

He revealed misgivings, however, at a meeting called by Teres, who had a grant to follow 209 terminal patients to evaluate medical approaches to dying. Teres asked if anyone could remember the first time he had seen a brain-dead patient. Higgins said he did: "It was uncomfortable to make the decision, go through the routine, and do all the diagnostic testing, and then conclude at the end of the exam that this patient was somehow different than they were before. Prior to that he was our patient for whom we were trying to do everything. After brain-death evaluation, it was a matter of turning off the ventilator, and I recall that as being very, very uncomfortable the first time. It has become progressively less uncomfortable over the years, and I've done dozens of these now. I feel an awesome responsibility when I make that evaluation."

Teres and Higgins agreed on the most common conditions leading to brain death: head trauma, cardiac arrest, cerebral trauma (e.g., aneurysm), drug overdose, and drowning. Teres says it's often confusing for the family. "They say, 'You're telling me that yesterday he was okay and today he looks just the same, only he didn't move when he was suctioned, and so he's dead?' " Teres says that sometimes to convince the family that the patient is dead—he gives the example of a Hispanic family—he brings them into the ICU to watch the tests. About one relative he said, "I had her sitting in the room while I did the brain-death evaluation. With each test I told her what a normal person would do. When they see that the patient doesn't react to the suctioning, ice water, et cetera, they're convinced. They will not sign the [organ donation] consent form unless you make it very clear to them the family member is dead."

Teres, Higgins, and their whole crew, as mentioned, appeared to be rigorous, serious, and respectful of patients. But Teres did say that though the policy on brain death is very clear, the death declaration business is not one that strives for perfection. "If you wait for everything to be a hundred percent, you'd never have organ donation."

CARBON DIOXIDE

James Bernat is a solid proponent of brain death, but he has concerns. A professor of neurology at Dartmouth Medical School, Bernat was on the Vatican task force that deliberated over whether *morte cerebrale* (brain death) should be sanctioned by the Pontifical Academy of Sciences. The majority of the panel, including Bernat, voted in favor of brain death. (The Vatican has vacillated on the issue a number of times since then.) He admitted that there had been some strong opposition. Bernat is one of the few proponents who examines the flaws in brain death.

Bernat says the problem is the application of the medical standards as opposed to the standards themselves. After a study of brain-death determinations, he estimated that two-thirds of them (65 percent, to be precise) are performed improperly. The primary problem is that most doctors, unlike the Baystate team, don't adequately monitor carbon dioxide levels in the patient's blood. The apnea test then becomes meaningless. No matter how alive the patient may be, he will have no impetus to breathe when the ventilator is disconnected. This is a stunning figure: 65 percent of brain-death declarations are done incorrectly. It doesn't mean that all of those people were not brain dead, but at least some determinations had to be wrong.

Unlike other proponents, Bernat accepts that there have been legitimate cases of patients declared brain dead who have returned to life, but he blames this on sloppy testing: "In every case I've looked at of a person who comes back from brain death, the test was done improperly. I am unfamiliar with anyone with brain death who has come back on whom the tests have been done properly." This appears to be comforting to him. It might be less comforting to the patient.[23]

Other things give Bernat pause: brain-dead pregnant mothers; locked-in patients, who have severe brain-stem injury but whose cortices are unharmed and are perfectly conscious; patients who recover after brain death and continue to be treated. More about those problems later. For now, a more imminent problem looms.

LAMPPOST SCIENCE

The UDDA states that for brain death to be declared, "the entire brain must cease to function, irreversibly. The 'entire brain' includes the brain stem, as well as the neocortex."[24] This is known as the "whole-brain" definition, and proponents of brain death use the catchphrase "whole brain" frequently. It is comforting to hear that doctors wait until the cortex is dead in that it is the locus of our consciousness, our senses, our pleasure, and most important, our pain. No one wants to be buried while still conscious. No one wants to be autopsied or cut apart when pain and suffering are still possible.

However, those of you who know a little neurology will recognize that Higgins's set of tests on Fernanda, rigorous as they might have been, lend insights only into the functioning, or nonfunctioning, of the brain stem. None of the procedures shed light on the workings of her cortex. She could have been calculating the cross section of the bottom quark using Heisenberg's matrices, and no amount of ice water squirted into her ear would have detected it.

Two paragraphs after insisting on whole-brain death, the

UDDA goes on to state, "This Act is silent on acceptable diagnostic tests and medical procedures. It sets the general legal standard for determining death, but not the medical criteria for doing so. The medical profession remains free to formulate acceptable medical practices and to utilize new biomedical knowledge, diagnostic tests, and equipment."

In other words, the UDDA insists on whole-brain death, then takes it away by saying that the medical establishment can test for it any way it pleases. How many doctors and/or hospitals actually conduct so-called *confirmatory* tests that might provide insight into the workings of the cortex? The doctors we talked to, such as Ross, Teres, Higgins, and various anesthesiologists, rarely referred to EEGs and so on, though a few others claimed confirmatory tests were commonplace. No statistics were provided, however. Alan Shewmon says the age of the patient is important: "As a pediatric neurologist, my direct clinical experience is with children, and in the ICUs where I consult, they typically do confirmatory tests. The younger the patient, the more important are confirmatory tests. In the adult patient world, there has indeed been a move away from confirmatory tests, with total reliance on clinical criteria. My impression is that the percent of adult cases where confirmatory testing is done, and the criteria that determine which cases get confirmatory tests and which don't, vary widely from institution to institution across the country, and maybe even within institutions from physician to physician. I have a suspicion, but no data, also that this shift away from confirmatory tests has occurred more in the U.S. than other countries."[25]

The EEG criterion recommended by the Harvard committee is long gone. Unfortunately, the EEG is the only test discussed by the Harvard committee that even attempts to measure activity in the cortex, though it was never reliable in the first place. For one thing, a "flatline EEG" had no clear meaning, in part because many doctors took "flatline" to mean any signal with a low amplitude. Thus one study reported that "10 percent of all adults" had flatline EEGs.[26] Conversely, when I visited the Jell-O Gallery in Leroy, New York, a few years ago, one display was captioned, "March 17, 1993, technicians at St. Jerome hospital in Batavia [New York] test a bowl of lime Jell-O with an EEG machine and confirm the earlier testing by Dr. Adrian Upton that a bowl of wiggly Jell-O has brain waves identical to those of adult men and women." No data were supplied on the EEGs of other Jell-O flavors. Unfortunately, the curator of the Jell-O Gallery could not verify the above citations.

The Jell-O experiment, if valid, would please the advocates of minimal brain-death testing. It would add credibility to the claim that EEGs obtained from otherwise brain-dead patients are simply "artifacts," the favorite word for such brain waves. They are using "artifact" not in the anthropological sense, "made by humans," but in the technological sense: the brain waves are being produced by the machine, not the patient. There are two sides to this argument, however. If the machine can produce brain waves to make a dead person seem alive, it can also fail to produce brain waves and make a living person seem dead.

The "whole-brain" standard for brain death is at the very least misleading—a construct, a fiction. Or, as we laymen might put it, a lie. It is lamppost science to look for life in the neocortex by testing the brain stem, after the old joke: a drunk loses his car keys in a dark alley, but he looks for them under a lamppost because, he says, the light is better there.

"Controversies in the Determination of Death," a white paper by the President's Council on Bioethics, published in 2008, tried to negotiate around the misnomer of "whole-brain death." The council admitted not only that the standard policy was to test only the brain stem, but that activity in other brain tissue, such as the pituitary gland, had been found in patients declared brain dead. The council did some verbal tap dancing to explain that "whole-brain death" did not have to include the "whole brain." That would be true only, it wrote, "if taken literally." The council goes even further, writing that "all functions of the entire brain do not have to be extinguished in order to meet the neurological standard," admitting that this "is somewhat at odds with the exact wording of the UDDA."[27] In other words, the council would say that "the whole nine yards" equals maybe two or three yards.

There is no doubt that the brain stem is important. We need it to function in order to breathe. Though we think of the cortex as the seat of consciousness, as "who we are," the brain stem's reticular activating system is essential for maintaining wakefulness, an essential ingredient of consciousness. Still, why is brain death based on this stalk of an organ, sprouting downward from the much larger cerebellum, which squats beneath the giant, by comparison, cortex? Visualize a pumpkin turned upside down. The brain stem would be the pumpkin stem. The meat of the pumpkin would be everything else.

The reasoning goes that the brain stem is the most resilient part of the brain. The white paper states, "if a brain injury has progressed to the point at which the brain stem retains no function, it has *probably* [emphasis mine] ravaged the more fragile parts of the brain as well . . . the so-called 'higher centers.' "[28]

Thus, if your loved one is being wheeled off to have his organs removed, he is *probably* dead. He is *probably* not conscious. Not all advocates of brain death subscribe to this "don't look, don't tell" policy. Bernat, for one, has recommended that tests for intracranial blood flow be added to the exam.[29] In Sweden, doctors must find a flat EEG to declare brain death and must also confirm that there is no cerebral blood circulation.[30]

IS THE BRAIN REALLY IN CHARGE?

Just as some of our ancestors saw the heart as the locus of the soul, today the medical establishment assumes that the brain is the organizing organ of the body, the "soul" as some doctors actually said, or, more commonly, the command center of what they call "personhood."

Dr. Alan Shewmon is the Antichrist of brain death. A pediatric neurologist at UCLA, Shewmon was a colleague of Bernat on the Vatican task force. Originally pro brain death, Shewmon has turned 180 degrees. Bernat hinted to me that a recent religious conversion by Shewmon had had a lot to do with his about-face. Shewmon laughed at the accusation. He said yes, he had converted "from atheism to theism," but the conversion had taken place in 1967 and had nothing to do with his scientific views.[31] The most scientific approach one can take to death, he said, is to treat human beings as any other species. People should be judged biologically on whether they are alive or dead, not on some vague notion of "personhood." There is no concept of "squirrelness," for example, or "gorillahood," by which we determine the death of other species. There we go on biological criteria alone.

Toward that end, Shewmon has become the nemesis of the medical death determination establishment. He has compiled 150 documented cases of brain-dead patients whose hearts continued to beat, and whose bodies did not disintegrate, past one week's time. In one remarkable case, the patient went twenty years after brain death before succumbing to cardiac arrest.[32]

Brain-death advocates have always insisted that anyone who meets their criteria will fall apart quickly and will soon meet the cardiopulmonary criteria.

That the brain is the body's integrative organ lies at the heart, so to speak, of brain death's alleged validity. Bernat, with coauthors Charles Culver and Bernard Gert, wrote a pivotal paper on the subject, in which they said, "the brain is necessary for the functioning of the organism as a whole." It integrates and controls bodily activities. Bernat et al. went on to write, "A patient on a ventilator with a totally destroyed brain is merely a group of artificially maintained subsystems since the organism as a whole has ceased to function."[33]

Shewmon counters this claim by arguing that patients who are diagnosed with brain death continue to exhibit many functions that one can call "somatically integrative," such as hemodynamic stability and body temperature, the elimination of wastes, an immune response to infection, and—most dramatically—a stress response when the transplant surgeon cuts into the patient's body to remove organs.[34] Shewmon insists that integration is an "emergent property" of the whole organism; it doesn't depend on any one organ or part.

Shewmon has compiled a list of life processes that his long list of brain-dead patients continues to exhibit:

- Cellular wastes continue to be eliminated, detoxified, and recycled.
- Body temperature is maintained, though at a lower than normal temperature and with the help of blankets.
- Wounds heal.
- Infections are fought by the body.

- Infections produce fever.
- Organs and tissues continue to function.
- Brain-dead pregnant women can gestate a fetus.
- Brain-dead children mature sexually and grow proportionately.[35]

Bernat and his coauthor, Eelco Wijdicks, dismissed Shewmon's cases as "anecdotes yearning for a denominator."[36] The denominator would be the total number of brain-death declarations, with Shewmon's cases in the numerator, which would give us a percentage of erroneous brain-death declarations. The total number of brain-death declarations in the United States is unknown. Bernat and Wijdicks obviously feel that since Shewmon cannot demonstrate how large a proportion of all brain deaths those 150 of Shewmon's cases comprise, it doesn't matter. If the total number is in the tens of thousands, perhaps Bernat and Wijdicks would feel that 150 mistakes is acceptable.

Patients, on the other hand, don't care about the "successful" brain deaths. They just don't want to end up as one of the 150 mistakes. The other argument for the validity of Shewmon's 150 cases is that in conventional science, all you need is one case to falsify a theory. If Bernat and other advocates were to say that *on average,* brain death is a good, or at least workable, standard, then the "anecdote without a denominator" argument might carry some weight. But they are not saying that. They are saying that brain death *is* death, and thus even one errant case destroys that theory.

Shewmon points out that the total number of brain deaths, even if we knew the figure (most guesstimate around 1 percent of all deaths), would be irrelevant. "The most relevant denominator would be the number of brain death cases where there was occasion to keep the patients going as long as possible [after brain death is declared]."[37] In most cases, no such effort is made, for obvious reasons.

DID SOMEONE MENTION TRANSPLANTS?

The question remains: why do we need brain death? Even advocates of brain death say that it quickly follows cardiopulmonary death. If, as its proponents say, brain-death criteria describe the same condition—that is, death—as the cardiopulmonary criteria, why bother? Especially since tools are available for declaring cardiopulmonary death and are sorely lacking, or at least ignored, for determining whether the whole brain is really dead.

One place to start is motivation. What drove the Harvard ad hoc committee to turn back the calendar and construct a lower standard for death? Motivation is important in science. Were the researchers open to the truth, in search of the facts no matter where they might lead? Or were they merely "curve fitting"? That is, did they have a theory already in mind and then cherry-pick their research, such as it was, to fit that theory?

To those of us with untrained eyes, it appears that the Harvard committee was fixated on freeing up organs for transplant by pulling the plugs on (and then replugging) patients on respirators. The Harvard ad hoc committee's report mentions organ transplantation in the very first paragraph. Brain-death advocates, on the other hand, defend the scientific integrity of the committee's work.

Unfortunately, Harvard University has closed the records of the committee for a fifty-year period. They are preserved with the records of the committee's chairman, Henry K. Beecher, at the Francis A. Countway Library of Medicine at Harvard Medical

School. The Dean's Office has opened the records to certain people, most notably Eelco Wijdicks in the Department of Neurology and the Neurosciences Intensive Care Unit, Saint Marys Hospital, Mayo Clinic, in Rochester, Minnesota, and Gary Belkin, at the time a clinical assistant in psychiatry at Massachusetts General Hospital and assistant professor of psychiatry at Harvard Medical School.

Wijdicks's article, which appeared in the journal *Neurology* in 2003, is the more factual of the two. Wijdicks writes up front that the 1968 Harvard committee has been "a target of criticism"—unfairly, he implies. The workings of the ad hoc committee "have not been carefully evaluated." Given that Harvard opened the files for him, he was in a position to do that careful evaluation. Oddly, most of the evidence Wijdicks presents appears to justify the criticism. First, let's briefly introduce the thirteen men on the committee, none of whom signed the historic article originally published in *JAMA*:

Henry K. Beecher (chairman), anesthesiology
Raymond D. Adams, neurology
A. Clifford Barger, physiology
William J. Curran, law
Derek Denny-Brown, neurology
Dana Farnsworth, psychiatry
Jordi Folch-Pi, neurochemistry
Everett I. Mendelsohn, history
John P. Merrill, nephrology
Joseph Murray, surgery
Ralph Potter, theology
Robert Schwab, neurology
William Sweet, neurosurgery

At least ten of the thirteen members were MDs. There may have been eleven. Ralph Potter, a theologian, was listed in one documentary as having an MD in addition to a ThD.[38] Neither Wijdicks nor Belkin mentions this. There is an abundance of doctors of the "neuro" variety (five), plus a psychiatrist and a theologian. In other words, seven were trained in matters of the brain/mind/nervous system. Two of the committee members, Murray and Merrill, were pioneers in organ transplantation. So there were nine members with philosophical and professional motives for wanting the brain to be the integrative organ of choice. Note the absence of cardiologists or pulmonologists to offset the brain-oriented doctors and to speak for the power of cardiopulmonary-death criteria. Since the committee was making the case that both brain-death and cardiopulmonary-death criteria are describing the same condition, death, would it not have made sense to include those specialists in the discussion?

Wijdicks concludes, in the opening paragraph of his article, that the Harvard committee acted "without special interest." In fairness, he presents the opposing view of critics, one of whom asked, "Is there an objective reality to be unearthed or is brain death a construct used by consensus to designate a patient as being dead?"[39] Two other ethicists felt the Harvard committee was being purely utilitarian. Disconnecting respirators would keep doctors from being arrested, and organs could then be gathered without violating the dead-donor rule.[40]

Wijdicks's defense of the committee's motives paradoxically includes one smoking gun after another. He begins by describing the medical milieu of the 1960s. The prime mover of the committee—more than even its chairman, Henry Beecher—was Joseph Murray, a surgeon who would go on to win the Nobel Prize in Physiology or Medicine for his triumphs transplanting organs. In his autobiography, Murray wrote about the fervor he witnessed at a kidney conference in those pioneering days of transplantation: "It seemed everyone wanted a piece of the action, and the attendees—mostly young, aggressive, and ambitious doctors—jockeyed for prominent positions in the field."[41] In most cases, donors in those days were patients whose hearts had stopped. In such patients, kidney function was deteriorating quickly.

Murray says he was alarmed by a doctor at the kidney conference who said, "I am not going to wait for the medical examiner to declare the patient dead. I'm just going to take the organ."[42] In addition to reading the papers and correspondence of the Harvard committee, Wijdicks also interviewed the three surviving members, Adams, Potter, and Murray. In 2003 Murray told Wijdicks that he had wanted to come up with some kind of criteria so he could protect the image of the profession.[43] Whether Murray had any concern for the patients, Widjicks did not say.

Beecher agreed with Murray about the purpose of the committee after a preliminary meeting: "I thought one of the few really positive products of the meeting was your contribution." Murray had two concerns. He wrote, "First is the dying patient, and the second, distinct and *unrelated* [emphasis mine], is the need for organs for transplantation."[44] Wijdicks apparently believed the two issues are unrelated because Murray said they were. Murray went on to say that to solve problem number one, brain death had to be installed as the new standard. When that is done, he continued, problem number two is immediately solved. Lots of organs will be freed up, and "all that is now required is proper permission from either the patient or the next of kin." Yet somehow the two issues, in Murray's words, were "unrelated." It is like a committee of foxes saying that it has two goals: (1) to make chicken coops with big holes in the sides, and (2) to meet the unrelated desire to eat more chickens. Murray punctuated his thoughts to Beecher with this amazing statement: "Can society afford to lose organs that are now being buried?" Patients, he said, are "stacked up in every hospital in Boston and all over the world waiting for suitable donor kidneys."

From April 11 through June 25, 1968, the committee went through six drafts of its article. One wonders if Wijdicks carefully read the published article, as he writes that the sixth and final draft does not actually mention transplantation. By my count, it does so twice. Wijdicks states that the correspondence "suggests that there was appropriate sensitivity not to link this work to transplantation."

The other transplant doctor on the committee, John Merrill, made a suggestion in writing to chairman Beecher. He said that "the decision to turn off the respirator should not include an opinion by a member of the transplant team." Merrill also suggested an end run around his own, and the committee's, desire to keep a barrier between the brain-death team of doctors and the transplant team.

> It would be ideal from the standpoint of both groups involved if some minor member of the team caring for the comatose patient, such as a nurse who was not involved in the decision making, could be aware that the patient was a good potential donor. The decision would then be made by the physician in charge but the nurse, being aware that he [the patient] was being considered as a transplant donor, might then simply request the act of turning off the respirator be postponed for an hour or so while the recipient was being readied. This, of course, would be a great help to the transplant team, and the

> decision would have been made by the physician and physicians in charge quite independently.[45]

What he's referring to here is that when the patient is declared brain dead, the transplant team will want the respirator reconnected immediately so that the "dead man" becomes a "beating-heart cadaver," his organs receiving oxygenated blood, keeping them viable for transplant.

Today this is done routinely. As for the idea of a "ringer," a nurse who is not a decision maker but who looks out for the needs of the organ bank and the transplant team, you could see that in action in the Baystate team that attended Fernanda. The nurse, Dennis O., was clearly keeping the OPO informed long before Fernanda was declared dead.

There was also a discussion among the committee members on the optimum time period between brain-death tests to satisfy both legal and organ transplant needs. Neurologist Raymond Adams objected in a letter to Beecher about using "the need of donor organs as a valid argument" for brain death. Theologian Ralph Potter was startled at the short period of time in which such an important document was completed. "It was not a deliberative body," he told Wijdicks.[46]

Despite the considerable discussion about transplants, Wijdicks concludes, "I am uncertain after reading the documents whether an alleged agenda of facilitating transplantation through a new construct of death existed. The interviewed members of the committee denied that transplantation was implicit in their deliberations."[47]

USELESS

Wijdicks did not include in his investigation the interviews committee members had given to others. Of particular note are neurosurgeon William Sweet and chairman Beecher, who were caught on camera by Sandoz, a pharmaceutical company that makes immunosuppressive drugs, for a film promoting the validity of brain death. The fiction that the committee had the dying patients and their families' interests at heart is dispelled when Sweet proclaims, "Once the state of coma becomes irreversible, we've got to stop this damn nonsense of blowing air into the lungs. We could save another life with them." Chairman Beecher says, "When coma becomes prolonged, the patient is useless to himself and others." Sweet echoes Beecher: "Practically, we could save another life with the organs of this individual who has for all intents and purposes passed beyond any state of utility ever again to himself or anyone else."

Further statements by Sweet indicate that Harvard may have erred in not including a few cardiologists on the committee. Sweet argued that the existing criterion of a stopped heart "is wrong. The heart is just a pump." That isn't true, of course. Sandoz got another expert, neurosurgeon Peter Black, to clear up the matter on camera. He said, "We're built in a very peculiar way. We need to have the brain to breathe. We don't need to have the brain to have the heart beat." The heart has a "brain" of its own. It is quite a bit more than a pump.

HARVESTING LIVE BABIES

Dr. Murray had also been interviewed elsewhere prior to Wijdicks's paper, in the *American Journal of Transplantation.* Murray performed the first successful kidney transplant in 1954, and in an era before immunosuppressive drugs, both donor and recipient were identical twins; they had to be to avoid organ rejection. Murray would win the Nobel Prize in 1990 for his work on organ and tissue transplantation.

For more than forty years, Murray has sat in the same pew every Sunday morning for 7:30 Mass at St. Paul's Church in Wellesley, Massachusetts, but his kidney operation changed the moral underpinnings of physicians forever. As a colleague noted, the doctor's credo "to first do no harm" became a thing of the past. It is difficult to say that one does no harm to a living kidney donor, slicing open his back and removing an organ. Prior to the milestone operation, performed at Peter Bent Brigham Hospital in Boston, "The decision to proceed was only given after the community was completely engaged, including discussions with prominent clergy of all faiths and legal scholars," wrote Francis L. Delmonico, Murray's interviewer and a professor of surgery at Harvard Medical School.[48] No such discussions took place prior to the formulation of the Harvard criteria.

Murray did not come off sympathetic to organ donors or their families. He said that operating on a brain-dead donor was for him "the start of an autopsy." He also told Delmonico that the Harvard criteria were too conservative. Murray would have included as dead those patients whose cortices were down but whose brain stems were still functioning. In other words, he would excise the organs from patients who could breathe on their own without the use of a ventilator.[49] Most stunning, Murray had no objection to the removal of organs from an anencephalic baby.[50] An anencephalic baby is one who is born without a forebrain, the cerebrum, and so on, but has a functioning brain stem. Awareness, if not consciousness as we know it, has been observed in anencephalic babies, and they are not considered dead by the Harvard criteria. Essentially, Murray, who was a prime mover on the Harvard committee, advocates taking organs from live babies.[51]

HARVARD ON HARVARD ON HARVARD

Gary S. Belkin's paper on the Harvard ad hoc committee, unlike Wijdicks's succinct, clearly written report, is a prolix, off-topic, meandering exercise in bureaucratese, in which Belkin seems to have taken little advantage of his special access to the correspondence and other papers of the committee. Though breathtakingly long, Belkin's "Brain Death and the Historical Understanding of Bioethics" rarely cites the discussions, deliberations, or letters of the Harvard committee. Instead, Belkin treats us to long abstruse passages of his own medical philosophy.

Oddly, Belkin writes that the committee's final report "was the product of men who knew and influenced each other." This is not comforting to many of us—that the definition of death was being reengineered by a baker's dozen of homogeneous buddies. In the end, Belkin clears the committee of having any desire to facilitate transplants. He also takes bioethicists to task for being critical of the Harvard committee and says that it was the committee members who were really the ethical ones by ending intrusive techniques for keeping patients alive "except as an attempt to harvest organs for others." Organ transplantation was not a motivating factor for the Harvard criteria, Belkin writes, but it "redeemed medical treatment in those cases."[52] If I understand Belkin's convoluted logic, the use of ventilators to keep "useless" patients alive was immoral. The Harvard committee redeemed this technology, however, by figuring out how to use it to harvest more organs. How the committee did this without thinking about transplants, Belkin does not explain. As noted, at the time of his report, in 2003, he was an assistant professor of psychiatry at Harvard Medical School. Thus Harvard opened its records to a Harvard professor to critique the Harvard criteria. He found the Harvard ad hoc committee ethically blameless. What are the odds?

CONFUSION AND FEAR

Despite the heroic efforts to clarify and justify the Harvard criteria, they remain opaque, confusing, and contradictory. Nurses sometimes write down the time of death as when the organs are harvested or, if the patient is not a donor, when his heart and lungs eventually stop functioning. The public is confused. A doctor said when he told one mother that her son was brain dead, she said it gave her hope. Families often think that brain death is a prognosis rather than a diagnosis or a declaration of death.

Doctors were also confused. As late as 1984, after the UDDA had been in effect for three years, a survey of neurosurgeons and neurologists showed that they were not united in their methods for declaring brain death. For example, 12 percent did not test for the absence of a corneal reflex, 16 percent did not test for doll's eyes, 41 percent did not test for dilated pupils, and 39 percent did not check for a cough reflex. More disturbing, 16 percent did not perform the apnea test. There are many conditions that can mimic brain death, including low body temperature (below 90 degrees) and barbiturates in the blood. Yet 44 percent did not take the patient's temperature, and 57 percent did not test for barbiturates.[53]

The committee revealed some insecurity. William Sweet recalled that a fellow member said, "What if a doctor says, 'This patient of mine met these criteria, and he's now walking around a healthy man today because of what I did for him?' "

"Well, that would be a devastating blow to the whole concept," said Sweet. "As far as I know, there has never been a case reported of an individual who met the criteria of that original Harvard committee who survived to tell the tale." He then gave the camera a little smile.

One wonders what the smile was about. One can only speculate. Perhaps it was confidence. Where many brain-dead patients go, to the organ harvest, there will be no worry about their ever walking or talking again. But there have been cases of people surviving brain death. Brain-death forces pooh-pooh them all, even those written up in medical journals. So I will choose one case that is hard to argue with, given that it is contained in the "bible of brain death," Julius Korein's *Brain Death: Interrelated Medical and Social Issues.* A patient was declared brain dead at a hospital in Pennsylvania. After donation consents were signed by the parents, doctors found he wasn't really dead and resuscitated him. Said one doctor, "It took us two weeks to get him sitting up in a wheelchair, eating, talking, and then after another two months discharged from the hospital."[54] The aftereffects were minimal.[55]

FOUR

The New Undead

They pee, they have heart attacks and bedsores. They have babies. They may even feel pain. They are "mostly dead." Meet the beating-heart cadavers.

I WAS scheduled to meet James McCabe in the lobby of Cooley Dickinson Hospital in Northampton, Massachusetts. A door marked "No Admittance: Hospital Personnel Only" suddenly flew open and out stepped McCabe, dressed entirely in white. Bright white pants, bright white shirt. On the street, he might be a Good Humor man. Or one might mistake him for a hospital employee, which I did, but he isn't. McCabe is the senior donation coordinator for the New England Organ Bank, its top organ wrangler. One suspects that his wardrobe choices were not random. Evolutionary biologists call this "crypsis," camouflage to help an individual blend in with his surroundings. A nonemployee emerging from a door marked "Authorized Personnel Only" said it all.

I conducted the interview in the hospital coffee shop, where we were shadowed by piles of doughnuts and cake domes encasing gooey baked goods on the counter. The coffee shop appeared to traffic in food items guaranteed to boost patient intake. Cooley Dickinson is one of those hospitals demonstrating that the primary stated goal of the Harvard ad hoc committee—to free up hospital beds—is outdated. Cooley Dick, as it's called in these parts, runs commercials on television making the hospital seem as inviting as a Club Med, a lumbar puncture like a piña colada sipped on the beach. Clearly there are empty beds to fill. One almost wishes that one were sick enough to take advantage of this comfortable hospital and its doughnuts. We sat at a centrally located table, and our conversation was interrupted on several occasions by doctors, nurses, and technicians saying hello to McCabe. Even if he wasn't an employee there, he certainly had the status of one.

I asked him about organ harvests. "First thing," he said, "get rid of the word 'harvest.' Bad word! Ten years ago a family ripped up a consent form [for organ donation] because an intern used the word 'harvest.' The uncle said, 'Harvest this!'" McCabe held up his middle finger. He said I should use "procurement," "retrieval," or "recovery." The last two words are most commonly used and grammatically make little sense. "Retrieve" comes from the old French root *trouver,* to find. Thus it means "to find again" or "to get back; recover; restore; rescue." Same deal with "recover." My kidneys are in the back of my torso. I don't consider them lost. That's precisely where I want them to be. Same with my other organs. In my mind, removing them does not restore or rescue them. To the organ transplant industry, evidently, my organs are all in the wrong place. They must be removed, put into an Igloo cooler with dry ice, and flown to somebody else who wants them. Only then will they be "recovered." For the sake of this book, we will use "harvest" to mean the extraction of organs from their original owners.

You'll excuse me if I at times lapse into the Orwellian vernacular of organ transplantation, which is too new yet to have developed its own vocabulary, so it just misuses otherwise useful English words. For example, "life support" is continued after a patient is declared brain dead if that patient is an organ donor. Dead donors who suffer heart attacks after being declared dead but before their organs are harvested are "resuscitated," meaning "revitalized; brought back to life."

McCabe was happy because yet another federal law favorable to OPOs had been passed, compelling hospitals to alert organ banks when a dead body was available. "I can now go to the hospital with a little bit of clout." He also said he could help hospitals muddle through the tangle of regulations and paperwork. Having seen the willing compliance of the Baystate brain-death team with an OPO and the compliance of doctors in general with the wishes of the organ transplant industry, more clout hardly seemed necessary. McCabe admitted that the OPO had embedded itself among hospital personnel. He said that nurses especially were very open to transplants and sent him many referrals. We saw that behavior with Dennis, the Baystate nurse, who was keeping the OPO informed even before Fernanda had failed the apnea test. McCabe felt he needed to educate trauma surgeons and neurosurgeons.

He goes to great lengths to get bodies. In one case the OPO had to bail the next of kin out of jail to get consent to take the patient's organs. His favorite donors are those with neurological failures, spontaneous bleeding in the head, high blood pressure, an aneurysm, or a head trauma—typically from a car or motorcycle accident. Head trauma accounts for 77 percent of brain-dead organ donors.[1] He once got the body of a chimney sweep who fell off a roof, a common hazard in that profession. McCabe prefers younger donors over older ones but says that a seventy-five-year-old kidney is as good as a seventeen-year-old kidney. One donor can provide many different organs, 3.3 on average, according to McCabe.

The transplant industry is a $20-billion-per-year business.[2] It spends more than a billion dollars a year on immunosuppressive drugs, which prevent the recipient's immune system from rejecting the transplanted organ. Transplant surgeons are near the top of the MD food chain. Earning on average around $400,000 per year, they and their staffs often fly to harvests on private jets. Finder's fees, in the form of "administrative costs," are often paid to hospitals.[3] The only people who don't get a share of the wealth are the most essential: the donors and their families. By law, they are the only ones who cannot be compensated. Joseph Murray, who performed the first solid-organ transplant—a kidney from one identical twin to another—and whose share of the Nobel Prize brought him $350,000, maintains that donors must not be paid in order to maintain the integrity of the field. I'm guessing that it also saves a lot of money.

It is the job of McCabe and other OPO wranglers to talk a family out of the organs of its soon-to-be-dead son, daughter, husband, wife, nephew, niece, or other relative. It must be one of the toughest sales jobs in the world. Distraught parents, say, whose child is dead or dying, are asked to make yet another sacrifice: the body of their son or daughter. How does one make sense of all this? That very pain and confusion, explains McCabe, is his way in. He offers the family *meaning.* "It's a way of finding meaning in death," he says. "Make the best of a tragic situation.

I'm going into the ICU to offer the family an option." The option most families want is to keep their loved one alive. A brain-death team tells them that's not in the cards. McCabe tells them the other option: to keep someone else alive. His batting average is excellent. He gets 50 to 60 percent of next of kin to agree to organ donation.[4]

Joanne Lynn, MD, a geriatrician and director of the Center to Improve Care of the Dying at George Washington University Medical Center, says, "Advocate groups just want the organs. Transplant debate has ignored the donors, and focused on the recipients. We're selling bereavement counseling."[5]

Whoo-hoo-hoo, look who knows so much. It just so happens that your friend here is only MOSTLY dead. There's a big difference between mostly dead and all dead. Mostly dead is slightly alive.

—Miracle Max, *The Princess Bride*

Organ transplants may seem peripheral to the story of death. They would be if they were what the organ trade claimed them to be: the neat extraction of body parts from totally dead, unfeeling corpses. It's more complicated than that. For the study of death, organ donation has been invaluable. It doesn't tell us so much what death is but what it is not. I present the following facts not as an attempt to derail organ transplantation—an impossible task, given how entrenched the industry is—but to take advantage of the knowledge, grisly as it may be, gained from our obsession with recycling the bodies of people who are, in the words of Pittsburgh's Michael DeVita, only "pretty dead."

If you want a more optimistic point of view, we are looking at the tenacious persistence of human life. Despite the fact that the Harvard ad hoc committee claimed that its criteria and the cardiopulmonary criteria described the same phenomenon, beating-heart cadavers (BHCs) are decidedly different from regular corpses. "I like my dead people cold, stiff, gray, and not breathing," says DeVita. "The brain dead are warm, pink, and breathing. They look sick, not dead." Beating-heart cadavers were created as a kind of subspecies designed specifically to keep organs fresh for their future owners. McCabe says that keeping the body alive from the brain stem down defeats warm ischemia, the restriction or loss of blood flow after conventional death. When the circulation stops, oxygen is no longer delivered to the organs, and cells begin to die.

McCabe says his outfit can get a donor from brain death to the operating room in twelve hours. Sometimes it may take an hour after death is declared to obtain consent, the ventilator being kept on while negotiations continue. An hour later a blood sample is drawn, and it takes eight hours to check for AIDS, hepatitis, and cancer, all of which disqualify a BHC from becoming a donor (with the exception of a primary brain tumor). Time is of the essence, because this brand-new creature—new since the advent of brain death—the beating-heart cadaver, or neomort, could easily have a heart attack and "die again" before his organs are removed.

McCabe's list of tests is only a partial one. At Baystate, Teres's team might do a bronchoscopy on the BHC to rule out pneumonia if somebody needs the lungs. Echocardiograms may be conducted for heart transplant. The lymph nodes are cut out and sent for tissue typing. Teres once took out a lymph node, put it in a taxi, and sent the cabbie from Springfield to Boston for typing. That's about a two-hundred-mile round-trip cab fare for a lymph node. Those of us among the living can only dream of such speedy test results.

Once a patient goes brain dead and relatives sign his organ

donation consent form, he will get the best medical care of his life. A hospital code blue may be a call for doctors to rush to the bedside of a beating-heart cadaver who needs his or her heart defibrillated. There are rumors of crash teams running into a room and reviving a BHC, while a patient in the adjoining bed has a DNR (do not resuscitate) sign on his chart. It's a popular legend, but I was not able to track down any such incident. It also seems unlikely in that BHCs can certainly afford a private room.

BHCs are also routinely turned in their beds so they don't get bedsores. Their kidneys are treated to avoid renal failure. Fluids are administered constantly to avoid diabetes insipidus, among other things; a healthy BHC should pee out 100 to 250 milliliters of urine per hour. If more, he has to be rehydrated. The lungs have to be monitored to keep them in shape for the next owner, and mucous is removed. Sometimes the brain dead are kept warm using aluminum foil blankets or warmed intravenous fluids or having them breathe gas that's been heated. There are debates over whether thyroid hormones should be administered.[6] It appears that the brain dead have adequate thyroid hormone. Cooper University Hospital's Steven Ross and Baystate's Dan Teres both say keeping BHCs "alive" is an arduous task for hospital nurses and other workers. Ross says it takes "very aggressive care." But that they can be medically cared for at all, as Alan Shewmon demonstrated with his 150 cases, gives one pause about the validity of their deaths.

THE HARVEST

There's more than one way to harvest a beating-heart cadaver. McCabe's outfit uses a team of seven in the operating room: one surgeon, one resident, one technician from the organ bank, one coordinator from the ICU, two nurses ("from the best hospitals," McCabe says), and one anesthesiologist. Some teams may add another surgeon if many organs are being extracted.

In a typical dissection, a long midline incision is made from just below the neck to the pubic area. The sternum is split with an electric saw or a Lebsche knife, a chisellike instrument the doctor hits with a mallet. A sternal retractor with spikes is used to open the sternum. Sometimes the aorta is clamped at the beginning of the harvest and the blood replaced with a fluid that McCabe says is "totally unique to the planet." It's a coolant, cold to the touch and clear but with a slightly yellowish tinge, like albumen or plasma. This avoids clots and stabilizes temperature.[7] Traditionally, the donor's blood is simply left in place.

There are many good, vivid descriptions of harvests in the popular literature. Some of the best are by three writers: Margaret Lock,[8] Mary Roach,[9] and Kathleen Stein.[10] Both Stein and Roach were impressed by the liver (as are many surgeons, the liver being one of the most difficult organs to transplant). Stein writes, "Aesthetically the liver is most pleasing, resembling some lustrous sea creature, smooth and supple, with sharply defined edges. But a surgical error contaminates it. As a result the liver loses its silkiness and definition, turning from coral pink to meat-market purple."[11] Roach writes, "But the liver gleams. It looks engineered and carefully wrought. Its flanks have a subtle curve, like the horizon seen from space."[12]

Lock, an anthropologist at McGill University, is less lyrical, capturing the gritty side of the business: "Now, in the operating room, the liver is exposed, but on seeing it the surgeon lets out an expletive. He asks someone to call a pathologist to determine whether the spots on its surface are due to cirrhosis, in which case the liver will be of no use. 'She only drank a little, did you say?' Turning to the transplant coordinator, the surgeon frowns behind his mask. 'Looks like more than a little to me.' "[13]

Michelle Au, an anesthesiologist, writes about working her first harvest at 3 a.m., remarking on how confusing it was because the donor "didn't *look* dead." (Emphasis hers.) After his kidneys were removed, she almost took the body back to the ICU. A nurse said no, take the body to the morgue. She also told Dr. Au that after working her first harvest, she went home and crossed off the organ donation box on her driver's license. She no longer wanted to donate. "It was a young guy, and they took everything. Heart, liver, pancreas, everything. . . . And after they all got what they wanted, everyone left, and the patient was just lying here alone."[14]

Dennis O., the nurse on the brain-death determination team at Baystate, appeared to be the most gung ho about using the organs of Fernanda. Long before Fernanda was officially dead, he was talking about the OPO. But even he had difficulty accepting that brain-dead people heading into a harvest are not real patients about to undergo real operations with real hazards. Dennis said, "When I went into a patient's room the other day he had been declared brain dead, and the organ donor people were here. I asked where his parents were, and they said after the brain-death declaration they left. I remember standing there thinking, intellectually I know he's brain dead, but if he were my son, would I have left before he went to the OR? Even working here you still have feelings about it."[15]

In the harvest she witnessed, Stein observed that the donor's heart suddenly accelerated from 100 to 200 beats per minute. The surgeons were alarmed, and shocked it back to normalcy with a jolt of electricity from defibrillating paddles.[16] Inquiring minds might ponder why a dead person's heart would suddenly become excited. It is not a question that interests the organ transplant industry. Lock writes that attitudes toward brain death and transplantation in Japan differ greatly from those in the United States. The Japanese press reports more aggressively on the imperfection of brain-death criteria, and transplant surgeons do not enjoy the high status afforded them in the United States. Some administrators throw them out of their hospitals.

At home, our anesthesiologists, whose primary duty is to spare their patients unbearable pain, are beginning to wonder about those racing heartbeats and other suspicious symptoms exhibited by donors. What does a "pretty dead" patient experience during a three- to five-hour harvest sans anesthetic?

SHACKLING THE OMBUDSMEN

Mark Schlesinger doesn't like his patients to feel pain during conventional surgery. He is chairman of the department of anesthesiology at Hackensack University Medical Center in New Jersey. In an operation, he says, the anesthesiologist is supposed to be the ombudsman for the patient. The surgeon is in charge of the operation. (There is a bit of truth in the saying, "The operation was a success, but the patient died." The patient is not entirely the surgeon's responsibility.)

Anesthesiology is more than just pain relief. Schlesinger says, "We must keep the heart working, maintain kidney function, et cetera. I might use ten drugs in one operation." Pain during surgery is a problem. The anesthesiologist doesn't want to overwhelm the patient with anesthetic, which makes recovery more difficult and creates side effects. At the extreme, an excess of anesthetic can push the patient into coma or kill him (the way veterinarians "put animals to sleep"). Too little, obviously, means the patient will feel pain during surgery. This happens, says Schlesinger, in as many as 1 percent of surgeries.

Thus the anesthesiologist keeps a close eye on the patient's vital signs. For example, says Schlesinger, you might see a patient's blood pressure spike. There are two ways to deal with the high pressure: either administer a drug that acts on it directly, such as heparin, or give the patient more anesthetic. The direct drug will solve the problem for the surgeon, who doesn't want to deal with high blood pressure during surgery, but it might just be covering up a problem: the high pressure may be caused by pain or awareness. So in a normal operation, says Schlesinger, the anesthesiologist will always try more anesthetic first, to see if that brings down the pressure. If that doesn't work, he knows that pain was not the problem and administers heparin or some other direct-acting drug such as Inderal or hydralazine to work on blood pressure or heart rate. I asked him how many times, in his experience, did more anesthetic bring down high blood pressure? "Every time," he said.

During a harvest, it's a different story. Technically, there is no patient, just a donor, who is dead, a nonperson. When the donor's blood pressure or heartbeat spikes, the anesthesiologist cannot act on his instincts and administer more analgesic. The surgeon, who is in charge, won't let him. The anesthetic could contaminate the organs, which have already been assigned to another owner. In a conventional operation, the patient's blood will clean the organs within several days. Clearly, that isn't the case after a harvest.

I asked Schlesinger why the transplant establishment doesn't allow a single anesthesiologist on a single harvest to administer anesthetic to combat increased blood pressure or heart rate—just to see what happens. If the symptom persists, pain would be ruled out as the underlying cause. But if anesthetic relieved the problem . . . Schlesinger considered it a stupid question, explaining that (1) transplant surgeons don't want to risk the organs, and, more important, (2) they don't want to know; finding that donors were in pain while their organs were being removed during a multihour operation would not be good for business.

Schlesinger points out that an anesthesiologist creates brain-dead patients every day: "We give drugs to make them die. And we bring them back [when the surgery is completed]." A patient under anesthesia is one of the many growing exceptions to the Harvard criteria. He would meet the criteria on the surface but would be disqualified (ruled still alive) if the examining doctor knew his system was full of drugs. "The only test you fail under anesthesia," says Schlesinger, "is irreversibility." That is, an anesthetized patient has had his brain stem put down temporarily. A brain-dead organ donor's brain stem is also down, but theoretically forever. But we don't know, given the limitations of the Harvard criteria and their focus entirely on the brain stem, what's going on with the donor's cerebral cortex or everything in his body below the brain stem.

John Neeld agrees with Schlesinger that the anesthesiologist must serve as the patient's ombudsman: "The surgeon is busy with his job. I'm going to render you unconscious. Paralyze all your muscles so you can't breathe. I'm going to breathe for you. Twenty-five to thirty million people per year come to hospitals and turn their lives over to people they don't know."

Neeld, of Atlanta, Georgia, is the chairman of the American Society of Anesthesiologists Association Section Council to the AMA. During surgery, he says, the anesthesiologist has four primary goals:

- To produce amnesia, using hypnotics
- To block pain, using analgesics

- To relax muscles so the surgeon can move them around
- To control reflexes such as the fight-or-flight response to keep blood pressure down

Toward these ends, Neeld uses hypnotics, such as pentathol, to put the patient to sleep and block his memory of the experience, muscle relaxants to relax the patient's trachea and put in a ventilator and endotracheal tube, and a general anesthetic such as nitrous oxide and desflurane.

Neeld is especially sensitive to patients who are aware or in pain during an operation. He was on call for surgery one day when his mother-in-law was brought into the OR with a broken hip. Neeld used only a light anesthetic. While operating, the surgeon said, "Hey, John, come look at this." Neeld said, "Is that what I'm afraid it is?" His mother-in-law had cancer. Later Neeld went to give her the bad news, but she already knew. "I heard you and Dr. Griffin talking," she said. He went back, looked at the operation's transcript, and found that her pulse had increased about the time of the conversation. Neeld disagrees slightly with Schlesinger's statistics, saying that .6 percent of patients are aware or in pain during surgery, not a full 1 percent.

To detect patients in pain, he looks to see if the patient is moving or sweating and watches for increased blood pressure or heart rate as the vessels constrict. He takes a careful look if he sees the surgeon tugging on a retractor, say, to get a look at the appendix. He agrees with Schlesinger: if you treat hypertension and other symptoms of pain directly, you mask the problem.

There are some surgeons, Neeld says, who want you just to keep the patient still, keep him from causing trouble. Even at the end of life, he says, "Every patient deserves humane treatment. I would hate to live with the knowledge that there was even a trillionth of a chance that the patient was in pain during the last moments of his life."

Neeld is an organ donor. "I believe in it strongly. But my first responsibility is to my patient. I can't go in [to an operation] with my hands tied." He believes that there are temptations. "'Dick, I'm going to do everything to keep you alive,' but in the back of my mind I'm saying, 'Hmmm, Dick's kidneys are still pretty good.'"

HARD TO IGNORE

Anesthesiologists have been at the forefront of questioning the finality of brain death and whether beating-heart cadavers are unfeeling, unaware corpses. They have also begun speaking out and writing about the subject.

Gail A. Van Norman, an assistant professor of anesthesiology at the University of Washington, cites three disturbing cases:

- An anesthesiologist administered a drug to a BHC to treat an episode of tachycardia during a harvest. The donor began to breathe spontaneously just as the surgeon removed his liver. The anesthesiologist reviewed the donor's chart and found that he had gasped at the end of an apnea test, but a neurosurgeon had declared him dead anyway.

- A thirty-year-old patient with severe head trauma was declared brain dead by two doctors. Preparations were made to excise his organs. The on-call anesthesiologist noted that the BHC was breathing spontaneously, but the declaring physicians said that because he was not going to recover, he could be declared dead. Besides, the proposed liver recipient would die without a new organ. The harvest proceeded over the objections of the anesthesiologist, who saw the donor move and react to the scalpel with hypertension that had to be treated. It was in vain since the proposed liver recipient died in another OR before he could get the organ, which went untransplanted.

- A young woman with pregnancy-induced hypertension suffered seizures several hours after delivering her baby. A neurologist said she had suffered a "catastrophic neurologic event," and she was readied for harvest. At that time the anesthesiologist found that she had small yet reactive pupils, weak corneal reflexes, and a weak gag reflex. After treatment by the anesthesiologist, "the patient coughed, grimaced, and moved all extremities." She regained consciousness, went home, and rejoined her family. She suffered significant neurologic deficits but was alert and oriented. Not dead.[17]

The brain-death establishment always discounts such cases as being "anecdotal" or undocumented as if they had all appeared in *The National Enquirer.* The above three cases appeared in *Anesthesiology,* the journal of the American Society of Anesthesiologists, which has 44,000 members.

The Harvard criteria state that the brain-dead patient must exhibit no movement. Van Norman, however, points out that some exhibit spinal automatism, or Lazarus sign, a complex spectrum of movements including flexion of limbs and trunk, stepping motions, grasping motions, and head turning.[18] Dr. Gregory Liptak, in *The Journal of the American Medical Association* (*JAMA*) writes, "Patients who are brain dead often have unusual spontaneous movements when they are disconnected from their ventilators. Numerous authors have described these disconcerting phenomena. Patients who fulfill all the criteria for death . . . may experience any of the following: goose bumps, shivering, extensor movements of the arms, rapid flexion of the elbows, elevation of the arms above the bed, crossing of the hands, reaching of the hands toward the neck, forced exhalation, and thoracic respiratory-like movements. . . . These complex sequential movements *are felt to be* [emphasis mine] release phenomena from the spinal cord including the upper cervical cord and do *not* [emphasis mine] mean that the patient is no longer brain dead."[19] According to the Harvard criteria, the dead patient should not be moving at all. Dr. Liptak's explanation and language—"are felt to be"—betray a decided lack of rigor in the brain-death business. In my years of covering particle physics, I never heard physicists say, "The particles in the accelerator *are felt to be* protons." They would want to check first and not rely on their feelings.

DOES IT HURT?

One cannot determine with certainty what organ donors feel, if anything, while being harvested. If transplant surgeons would allow what Schlesinger suggests—the administration of anesthetic when the donor shows normal signs of distress—we might discover the nature of a beating-heart cadaver's consciousness or at least gain some small clues. To reiterate, advocates for brain death claim that the brain is the organizing organ, and when it goes, what remains is, as James Bernat put it, "no longer a functional or *organic* unity, but merely a *mechanical* complex."[20] But pain responses and movement would seem to transcend "mechanical" reactions. The logic of brain death goes like this: if the brain stem is dead, the higher centers of the brain are also probably dead, and if the whole brain is dead, everything beneath the brain stem is no longer relevant. Since in practice only the brain stem is routinely tested, the vast majority of the body, everything above the brain stem and everything below, no longer counts as human.

The law more or less agrees. George Annas, a lawyer specializing in health law, says, "It's in or out. If you're alive you're a person; you're protected by the Constitution. If you're dead, our only obligation is to bury you. So it's very, very critical that we know who's in and who's out."[21] The law is on the side of the person with the scalpel. If a donor were to wake up in the middle of a harvest and demand a lawyer, the transplant team could ignore him and continue snipping out organs.

Giving painkillers to BHCs during a harvest is controversial. The reason for denying BHCs anesthetic during the removal of their organs is hard to pin down. Joanne Lynn and Schlesinger say it's because anesthetic will harm the organs. A critical care consultant in England, Tom E. Woodcock, rejects the claim that when the brain stem goes, the heart will stop beating soon thereafter and says it's time to allow families to choose that anesthesia be used for BHCs undergoing organ removal.[22] *The Canadian Journal of Anesthesia,* on the other hand, tells its subscribers that during organ removal "providing analgesia and unconsciousness is unnecessary." It also tells them to ignore muscle twitching and complex movements of the limbs and trunk and warns them in the typical Orwellian transplant language, "Diabetes insipidus is a common complication of brain death."[23] One would normally hope that after dying one could stop worrying about complications. Despite all this, administering anesthetics to BHCs during organ harvests is becoming more common in Europe, according to Robert Truog, a professor of medical ethics, anesthesiology, and pediatrics at Harvard Medical School and one of the most vociferous opponents of brain death.[24]

On three occasions during the past decade, I made attempts

to donate my organs through an OPO. I did not say I was a journalist; I was just a person wanting to give up his liver, kidneys, and so on. In each case I asked if I could arrange to receive anesthetic when my organs were being removed. The e-mails went back and forth with the OPO telling me I would be dead and dead people don't feel anything. I would reply, fine, but indulge me; I'll be happy to pay for the drugs, and I'll set aside a fund in my estate to make sure the hospital, OPO, and transplant team are paid adequately. Then the OPOs would come up with all sorts of procedural barriers, never getting to the point that perhaps they did not want to give me painkillers for the usual reasons: (1) there might be damage to the organs and (2) it would be an admission that a harvest is potentially painful.

My frustrating dealings with the OPOs reminded me of Jack Nicholson's character in the movie *Five Easy Pieces.* To overcome the OPO's procedural difficulties in giving me anesthetic, I felt like saying "Look, you've got a hospital. You've got anesthetic. You have an anesthesiologist. You have an operating room." Then I would commission a second operation, a face-lift, with a second surgeon and a second anesthesiologist. I'd get my anesthetic, and I'd hold the face-lift.

Eventually, I asked them point-blank to tell me whether they agreed with some anesthesiologists that there was the possibility of pain. At this point, all three OPOs broke off correspondence or avoided the question by giving me a definitional response: "You'll be dead, and dead people don't feel pain." See the endnotes for the unedited e-mail correspondence with the New England Organ Bank.[25]

There are few stronger opponents of brain death than Alan Shewmon and Robert Truog, yet both refuse to acknowledge the possibility that some donors may be in severe pain during organ harvest. Both got angry when I raised the pain controversy, but both acknowledged that donors did exhibit reactions to the scalpel and other activities, as did inadequately anesthetized living patients undergoing conventional surgery and who afterward reported pain and consciousness. Shewmon said it was simply "bodily reactions to noxious stimuli." Pain would fit nicely under that definition, I noted, to which Shewmon said that pain is "subjective." I asked if an experiment could be designed to answer the question of pain in donors. He said no. Shewmon goes on at greater length in the endnotes.[26]

Truog[27] was even angrier with me. At first, he didn't even want to discuss the topic of pain in the organ donor, comparing it to an argument over "whether it was okay to kick a rock."[28] I told him my tale of trying to sign up as a donor but asking for anesthetic, with no luck. Truog said that the reason for denying me anesthetic was not so much that the chemicals harmed the organs as it was a philosophical one: administering anesthetic would betray the uncertainty of the donor's alleged death. Again I suggested experiments along the lines suggested by other anesthesiologists: when BHCs show pain reactions during a harvest, administer anesthetic to see if the reactions subside.

Truog surprised me by saying he had already done so. He has used two different kinds of anesthetics, which will not harm organs, to quell symptoms such as high blood pressure and heart rate in harvestees. "Just because the symptoms come down, though," he added, "does not mean the patient is in pain. Pain is a subjective thing." As with Shewmon, I asked Truog if an experiment wasn't called for. Annoyed, he said no, that there was no experiment that could answer the question of pain in the donor.

I have been a science writer since 1973, and it was only the third time I had been told that a hypothesis could not be tested experimentally.[29] That donors feel no pain is not just a hypothesis but a working hypothesis, by which people are sent to harvests. That two major doctors say it cannot be tested via experiment signals that we have left the realm of science.

Truog unloaded another blockbuster, this one apparently from humanitarian motives. He suggested that if I sign up to be a donor or sign over a family member for donation, I should request two anesthetics: high-dose fentanyl and sufentanil. "If it were my family," Truog said, "I'd request them." That a pro-organ-transplant spokesperson wants anesthetics for his family speaks volumes. He added that his request, coming from a Harvard doctor, would have a good chance of being honored. Those of us in the hoi polloi, not so much.

Candace Pert, the neuroscientist who discovered the first receptor in the brain, the opiate receptor, takes issue with the assumption that death of the brain stem—which is what "whole-brain death" really means—translates to BHCs not being able to experience pain. She notes that the brain stem has nothing to do with pain. "Pain is an hallucination of your brain," she says, but it's the forebrain, the cortex. "The regions of the brain are well connected," she says. "That one area is dead is meaningless. Just because the brain stem is dead doesn't mean the cortex is dead."[30]

The alleged subjectivity of pain was nicely addressed in *Gourmet* magazine by David Foster Wallace in a 2004 article entitled "Consider the Lobster."[31] The science of organ retrieval is well mirrored in a "Test Your Lobster IQ" quiz published by the Maine Lobster Promotion Council. It states, "The nervous system of a lobster is very simple. . . . There is no cerebral cortex, which in humans is the area of the brain that gives the experience of pain." Wallace goes on to explain that the neurological facts militate against the no-pain theory of boiling lobsters alive.

He writes that lobsters have nociceptors, pain receptors sensitive to damaging extremes of temperature, "as well as invertebrate versions of the prostaglandins, and major neurotransmitters via which our own brains register pain."[32] Pain may be subjective, but Wallace points out that behavior is a clue. He describes dumping a live lobster from a container into a kettle of boiling water. The lobster clings to the container's side or hooks its claws over the kettle rim to keep from falling in. If you cover the kettle, you can hear the clanking of the lobster trying to push it off or the claws scraping the sides.[33] But, as with an organ donor, we can't for sure interpret what the lobster is experiencing. Then again, Wallace writes, some cooks can't take the sound of the crustacean's death throes and leave the kitchen while the lobster is boiling, taking an oven timer with them.

I do not mean to equate organ donors with lobsters being cooked alive. It is estimated that it takes a lobster between thirty-five and forty-five seconds to die in boiling water.[34] An organ harvest can take several hours. This might not discourage all donors. In a gathering of religious scientists, during which we discussed the possibility of pain during an organ harvest, one man, a Christian microbiologist, said he would welcome the pain. It would in fact provide more meaning to the gift of his organs, as he would undergo suffering equal to that of Jesus on the cross.[35]

Almost as uncomfortable as a lobster in hot water is Truog, who is trying to expand the pool of donors for his allies in the transplant game but is nevertheless logically bothered by the fiction of brain death and thus has made himself somewhat of an embarrassment to the take-an-organ-from-one-person-and-sell-it-to-another industry. "I am hurt that I am not invited to speak at the mainstream transplant conferences," says Truog.[36] Lynn

Margulis said, "Medical science is an oxymoron. It's not about finding the truth. It's all about please the teacher."[37]

"No one wants to rock the boat," says Truog, "including me."

The New England Organ Bank claims in its sales literature to donors that "there is no disfigurement associated with organ and tissue recovery." Hospital personnel and morticians disagree. In "long-bone retrieval," for instance, the thighbones are removed and sometimes replaced with broomsticks. A body without eyes is also an eerie sight. One organ donor coordinator, despite having logged hundreds of hours in ORs, says she still cannot watch eye removals.[38] Strips of tape are placed over the eyes of the BHC to keep the corneas moist throughout the long harvest. When all the other parts have been taken, an ophthalmic surgeon comes in to pluck out the eyes. Margaret Lock, who watched this, said, "I find myself repelled by this last intrusion in a way that I had not expected. For me, it seems, removal of the eyes represents more of a violation than does procurement of internal organs."[39]

THE MARGINALIZATION OF THE DONOR

Before there were beating-heart cadavers, there were organ transplants. In 1967, the year before the Harvard criteria were hatched, Christiaan Barnard performed the first heart transplant in South Africa. The donor was dead-dead, not just "pretty dead." Her heart had stopped. The recipient was Louis Washkansky, and the donor was a white woman named Denise Ann Darvall, the victim of a traffic accident. There was concern that the first donor not be "colored." Barnard declared her dead himself. Barnard remembers the dramatic moment, moving the heart from Darvall to Washkansky: "There was just an empty sac with no heart there, and it was really fantastic when we applied a shock to the transplanted heart, and the heart started to beat."[40] One wonders why Barnard couldn't have left the heart in Darvall and shocked it there. Washkansky lived eighteen days with the new heart.

Barnard's second heart transplant transferred the heart of twenty-four-year-old Clive Haupt, who was "colored," into the body of a white, Jewish retired dentist, Philip Blaiberg, aged fifty-eight. The dentist was asked if he had any objection to receiving a "colored heart." He said no. *Ebony* magazine, which had praised the first transplant, noted:

> If Dr. Blaiberg completely recovers and again walks the streets of Cape Town, a most ironic situation will ensue. Clive Haupt's heart will ride in the uncrowded train coaches marked "For Whites Only" instead of in the crowded ones reserved for blacks. It will pump extra hard to circulate the blood needed for a game of tennis where the only blacks are those who might pull heavy rollers to smooth the courts. It will enter fine restaurants, attend theaters and concerts and live in a decent home instead of in the tough slums where Haupt grew up. Haupt's heart will go literally to hundreds of places where Haupt himself could not go because his skin was a little darker than that of Blaiberg.[41]

Haupt's doctor, Raymond Hoffenberg, told *Twice Dead* author Margaret Lock in 1998 that the transplant surgeons were "hanging around" the ICU where he was caring for Haupt, and he had to send them away, insisting that Haupt was not yet dead.[42] Lock noted that early donors were lionized in the press as much as the recipients were. Washkansky's wife was shown looking sympathetically at Darvall's grieving father, who had lost both his daughter and his wife when they were hit by a speeding car. By 1969, however, Lock writes, "Donors became cloaked with anonymity as the ambiguity of their condition drew attention."[43] The American transplant surgeon Denton Cooley said, "In my opinion the clinician can become too preoccupied with the rights of the dead, namely the donor, at the expense of the recipient. We should not jeopardize the possible survival of the recipient while we are waiting around to make a decision whether the cadaver, as you call it, is dead or not."[44]

Blatant problems such as Barnard declaring the death of the donor whose heart he intended to transplant and surgeons loitering around Haupt as he fought for his life have been eliminated in word, but not necessarily in deed. We have seen how the Harvard ad hoc committee recommended behind closed doors that a member of the transplant team should not be allowed to declare brain death but suggested that a nurse or technician be recruited who was sympathetic to organ transplant to keep his eye on the progress of the testing and who would suggest leaving the ventilator connected to keep the organs perfused. That strategy has been followed with more zeal than the Harvard committee could have dreamt. Beyond nurses and others keeping the OPOs informed on which patients are likely to go brain dead, some doctors are concerned about "embedded coordinators" and trained "facilitators" who compromise the care in ICUs to ensure early identification of potential donors and put pressure on relatives resistant to organ donation.[45] One doctor, John Shea of Toronto, complains about "palliative caregivers" hired by the OPOs to make families feel comfortable. It's what he calls a "soft-sell program."[46] Michael DeVita worries that laws dictating that end-of-life care must not preclude organ donation put ICUs in the position of caring for organs rather than patients.[47]

OTHER TRICKS OF THE UNDEAD: SEXUAL AROUSAL AND GIVING BIRTH

I mentioned earlier that the much-ignored 1971 paper "Brain Death: A Clinical and Pathological Study" (aka the "Minnesota criteria") is full of surprises. The Minnesota team studied twenty-five moribund patients, conducting autopsies on them all and EEGs on some. They also checked for reflexes and found something unusual. Five of the twenty-five brain-dead people were still sexually responsive. The researchers gently stroked the "nipple and areola" of all patients and got responses from five, four men and one woman. Then they stroked the skin at the root of the penis of the eighteen male patients, and four responded with "gentle see-saw movements of the penis." The researchers felt this reaction was "an incomplete or abortive form of penile erection." Abortive or not, to the untrained eye it would appear to be a sign of life. Thirty million men in America have used erectile dysfunction drugs (Viagra, Levitra, Cialis), and 40 percent of them, or 12 million men, still get no relief from their impotence. They would probably be happy with a gentle seesaw movement.

More dramatic are brain-dead pregnant women. The first recorded case occurred in 1981 when a twenty-four-year-old woman, twenty-three weeks pregnant, was admitted to the State University of New York (SUNY) at Buffalo Hospital. After eighteen days her EEG showed no cerebral activity and she was transferred to the maternity division of the Women & Children's Hospital of Buffalo. A day later she was declared brain dead, approximately twenty-five weeks pregnant. So she was dead but still pregnant, and doctors decided to use her body—technically it was a corpse—as a meat incubator. It was not easy. She developed diabetes insipidus, sinus tachycardia, and uterine contractions. Later she had wide fluctuations in blood pressure, and the fetus's heart rate dropped. A cesarean section was performed immediately, delivering a two-pound "vigorous" baby girl at about the twenty-sixth week of gestation. Three months later she was discharged from the hospital, weighing about 4.4 pounds. Of note is that six months earlier another pregnant woman in desperate straits had been admitted to the same hospital.[48] The doctors had discontinued life support short of brain death as the fetus was nineteen weeks old and the medical staff accepted the belief that a body could not survive long after brain death was declared. There was theoretically not time to gestate the fetus another three weeks, twenty-two weeks being the earliest a viable baby can be delivered.

The second case ostensibly destroyed the theory that brain death is the same as heart/lung death, because the latter would follow within days if not hours of the declaration of brain death. These first two cases were published in *JAMA,* which also published in the same issue an editorial critical of brain death. *JAMA* noted that the authors of the brain-dead pregnant moms article "endorse the brain-death definition of death but repeatedly speak of the brain-dead mother as alive. For example they state that their goal was 'to prolong maternal life . . .' Their linguistic inconsistencies are, we believe, more than mere infelicities in usage. They reveal a deep—and in our view, justified—ambivalence."[49] You may recall that *JAMA* was the same journal that published the Harvard criteria article in 1968.

More brain-dead pregnant moms followed. At this writing there have been twenty-two published reports from around the world, including Brazil, Germany, Ireland, New Zealand, France, Finland, Korea, Spain, and the United States. From these twenty-two brain-dead mothers, twenty babies were produced, with no remarkable side effects in the infants. One woman gestated a fetus for 107 days after declaration of brain death.[50]

The concept of a dead but still pregnant woman has raised conflicts for nurses, among others. How should a nurse treat the brain-dead mother? As a living human being, a patient, or a corpse? The American Nurses Association acknowledges that "caring for the brain-dead physical body for long periods may be frightening or distasteful" and that the new technology "also impedes care by alienating and dehumanizing both the nurse and the patient." The organization concludes, "Thus the American Nurses Association would support the nurse who wished to consider the brain-dead person as a unique individual and worthy of nursing care. Yet a nurse would have to agree voluntarily to provide this type of care. . . . The nurse would have to find positive meaning in the actions. If a nurse feels that providing this care would create a value conflict, and that keeping a body alive for the sake of another is not right, conscientious objection would be appropriate."[51]

In other words, the Harvard criteria have put nurses into a catch-22 situation. The medical establishment, backed up by the UDDA, has ruled that the brain dead are not people. If a nurse agrees with this theory, she probably doesn't have the sensitivity to administer to brain-dead pregnant mothers, who are very needy from a nursing point of view. On the other hand, if a nurse finds this theory to be fallacious, she probably qualifies as a caregiver for brain-dead pregnant mothers. The association has not, to my knowledge, taken a similar stand on the nursing care of run-of-the-mill BHCs, those being kept on mechanical support for less than a day before their organs are harvested.

There are odd legal consequences of the technology. In some countries, the fetus is accorded legal protection at different stages of gestation: fourteen weeks in Austria, Belgium, Cambodia, France, Germany, and Romania; eighteen weeks in Sweden; twenty-four weeks in Singapore and the United States. There are generally exceptions, such as when pregnancy threatens the health of the mother. But in the above cases, the mother is legally *dead.* Though the mother is clearly being exploited without consent, she has no legal voice in the matter. And from the doctors' point of view, they may not have a choice either. Depending on the gestation period of the fetus and the country in which it resides, it might be tantamount to illegal abortion to disconnect the mother from mechanical support.[52]

The real significance of pregnant brain-dead women is that they would seem to sound the death knell for brain death. As Shewmon and many, many others have pointed out, what is more indicative of life than gestating a baby to a live and viable birth? Keeping a pregnant mother and baby "alive" for 107 days at the very least puts the lie to the theory that the brain dead will go quickly to conventional heart/lung death. At first, brain-death advocates said it was a matter of hours. Then they said three to five days at the most. Then seven days, then nine days, then fourteen days.[53] Now we're talking about a brain-dead mother not only hanging on for 107 days but nourishing a baby as well.

The only argument I've heard from brain-death defenders is that none of the pregnant mothers returned to a normal life. So even though one lasted 107 days past the declaration of death, she didn't drive home from the hospital. At least brain death has thus retained its status as a *predictor* of death. Jerome Posner, the coauthor with Fred Plum of the influential 1972 book *The Diagnosis of Stupor and Coma,* wrote in 2007 in the title of an article for the Proceedings of the Vatican Working Group that "Delivery of Live Babies from Brain-Dead Mothers Do[es] Not Negate Brain Death." Beyond the title, however, Posner did not give us satisfying details on why it does not. He cites a nebulous statistic: that at one transplant center, of 252 brain-dead women of childbearing age, only 7 were pregnant.[54] He doesn't explain this further. I guess that what he's saying is that not many people who go brain dead are pregnant. I'm not sure how that is relevant. His other point is that death "is a process." This seems to be a mantra in the field. Posner writes, "The physician can be certain that death has occurred, but cannot define exactly when." Indeed. That is the problem. In the case of one mother, the doctors were off by 107 days.

David Haig, a Harvard evolutionary biologist, takes a different tack. He has never considered the prenatal relationship of mother to fetus as being one of cooperation but closer to host-parasite relationships. When the mother's body doesn't make enough nutrients available, the placenta, which is the fetus's advocate, secretes various chemicals to increase blood flow and take more resources, at the mother's expense if necessary. When a pregnant woman has food cravings, she's most likely just filling orders phoned in by the placenta. Haig theorizes that preeclampsia is a disease caused by fetal-maternal conflict, with the placenta winning. His point: that we shouldn't make too much of the brain-dead mothers "supporting" their unborn children. The placenta is supporting them. The mother, alive or brain dead, is just a food source.[55]

A final note: brain-dead mothers can still donate their organs. Thus, after suffering a neurological catastrophe, being declared dead, still having to endure several weeks of pregnancy, then giving birth via cesarean section, the patient can still be rolled off to have her organs removed.

FIVE

Netherworlds

We believe they are unthinking, unfeeling. They wander the way stations between life and death, trying to tell their stories to an obtuse world.

IN 1997 Kate Bainbridge, a twenty-six-year-old schoolteacher living in Cambridge, England, came down with a virus. On the third day she sank into a coma as encephalitis ravaged her brain stem. Months later, in a nursing home, she wasn't responding to anything and was considered vegetative. Things might have remained that way had she not had the good fortune to come to the attention of a Cambridge neuroscientist, Adrian Owen, who put her into a positron emission tomography (PET) scanner. "We just thought we'd put her in the scanner and show her lots of bright colored images to see whether we could activate the visual cortex. . . . And what we saw was exactly what you would see in a healthy, awake person."[1]

Bainbridge, the first vegetative patient to have a PET scan, turned out to be far from vegetative. While she was in the scanner, she was shown a series of photographs of family members interspersed with meaningless scrambled images. When she saw the photos of her mother and father, an area of her brain known as V1 burst into activity. From scans of normal volunteers, V1 was known to be involved in facial recognition.

She does not remember being scanned, but on that day, she began a journey back from an unimaginable nightmare. Owen continued to scan her at intervals, and the scans showed that she was becoming steadily more aware.

For almost two years Bainbridge was paralyzed and unable to communicate in any way. She could not speak, move, hear properly, or swallow. She was fed by a feeding tube. She was, however, completely aware. Tormented by terrible thirst and racked with pain, she had no way to signal her caretakers. "I could not move my face, so I could not show how scared I was."[2] When she cried out in pain, the staff ignored her cries, dismissing them as reflexes. She was moved around as if she were an inanimate object.

"Not being able to communicate was awful—I felt trapped inside my body. I had loads of questions, like 'Where am I?,' 'Why am I here?,' 'What has happened?' They never even told me how they were feeding me. They said I couldn't feel pain but they were so wrong. I just couldn't tell anyone." Overcome by pain and despair, she tried to commit suicide by holding her breath.[3]

Then came the scan that gave hope to her family. Her nursing home brought in a neuropsychologist, Barbara Wilson, to assess her intellectual level. "I did a thorough assessment and it was obvious that Kate's intellectual functioning was normal. She didn't need help with her cognitive deficits because she didn't have any. But she was incredibly angry about how she'd been treated."[4]

Her recovery was painfully slow. It took her five months to recover the ability to smile. Eventually she regained some movement and became adept at communicating with a letter board, pointing to each letter in turn to say, "My story is about not giving up. I feed myself through a PEG tube. I have no sense of smell or taste."[5] She had been engaged to marry when she fell ill, but that life is over. Confined to a wheelchair, with multiple disabilities, she lives with her parents in South Cambridge, England, where she paints, reads and sends e-mail, listens to music CDs and books on tape, and occasionally goes to the cinema.

Based on her experience, she believes that "they should never let people die without assessing them properly." In her state she might have had her feeding tube withdrawn and been left to starve. If she'd been on a respirator and failed to react to having ice water poured into her ear, she might even have been mistaken for brain dead and had her organs harvested. "It really scares me to think what could have happened if I hadn't had the scan," she says.[6]

Not surprisingly, her case made a big splash. "No one had shown that a patient could activate [the visual cortex] totally normally while apparently in vegetative state," according to Dr. Owen. Bainbridge had been misdiagnosed, of course. She was not really vegetative; she just appeared that way. Such cases, in which patients return from the void with their intelligence intact, are relatively rare, but misdiagnosis is all too common.

There is dead, by which we normally mean dead in a cardiopulmonary sort of way, and then there is what Michael DeVita calls "pretty dead," or dead in a brain-stem sort of way. Then there are way stations, netherworlds that are nothing like everyday conscious life but that nevertheless place their occupants in the realm of the living and not the dead. In this chapter, we shall explore how easily the living can be mistaken for dead, how the conscious but paralyzed and silent can be mistaken for unconscious, and how most of us are prone to label as "better off dead" people whose conditions we know nothing about. We will be looking at some remarkable people who are in persistent vegetative state, minimally conscious state, locked-in syndrome, and coma.

The OPOs have their eyes on such people, especially those suffering from persistent vegetative state (PVS), of which there may be as many as 100,000 living in the United States.[7] There are many doctors who see these patients as a large new pool for organ donation. In Robin Cook's 1977 novel *Coma,* surgery patients undergo suspicious anesthesiology mishaps, become brain dead, and are stockpiled as waiting organ donors at the mysterious "Jefferson Institute." Cook was using "brain dead" in a general rather than a technical manner. Those patients were likely in PVS. Many doctors today would like to make the Jefferson Institute a legal reality. Cook was unfortunately all too prescient.

DESCARTES ALL OVER AGAIN

Let's travel back to 1972. That's the year "persistent vegetative state" was defined by Dr. Fred Plum of New York and Dr. Bryan Jennett of Glasgow to describe a case of "wakefulness without awareness." "In our view," the authors wrote, "the essential component of this syndrome is the absence of any adaptive response to the external environment, the absence of any evidence of a functioning mind which is either receiving or projecting information, in a patient who has long periods of wakefulness."[8] The poster child for PVS was Karen Ann Quinlan, who was very much on the scientists' minds and who, they said, "has not been aware of the extra years of life she has had, and thus has had no benefit from them." (Jennett is also the father of the Glasgow Coma Scale.)

According to this definition, the PVS patient is awake and can breathe on her own and maintain blood pressure but cannot swallow and must be fed through a feeding tube. Unaware of self and others, she is incapable of willful activity and does not feel sensation, including pain. In other words, the patient is an empty husk, a body without consciousness.

The definition is Cartesian—"I think, therefore I am." Plum and Jennett aspired to define an irrecoverable state called "permanent vegetative state" but in the end had to settle for the next best thing: persistent vegetative state. This is because a small number of patients defy the prognostications and recover (to some degree).

In 1994 the vegetative state was revisited by the Multi-Society Task Force on PVS, composed of representatives from many branches of medicine.[9] The task force determined that a vegetative state was distinguished by the following criteria:

> (1) no evidence of awareness of self or environment or an ability to interact with others; (2) no evidence of sustained, reproducible, purposeful, or voluntary behavioral responses to visual, auditory, tactile, or noxious stimuli; (3) no evidence of language comprehension or expression; (4) intermittent wakefulness manifested by the presence of sleep-wake cycles; (5) sufficiently preserved hypothalamic and brain-stem autonomic functions to permit survival with nursing or medical care;

> (6) bowel and bladder incontinence; (7) variably preserved cranial-nerve reflexes (papillary, oculophalic, corneal, vestibulo-ocular, and gag) and spinal reflexes.

In sum, a wakeful unconscious state that lasts longer than a few weeks is referred to as a vegetative state.

A vegetative state is distinguished from coma by the presence of sleep-wake cycles; from brain death primarily by the patient's ability to breathe on his own and maintain blood pressure and the presence of reflexes. If the patient's eyes track, or he is able to follow simple commands, if he says a word or reacts in a purposeful way to pain, touch, or words, he is considered to be in a minimally conscious state (MCS).

MCS results from less severe but generally widespread cerebral damage. Fragments of awareness remain, however. Patients may reach for objects, visually fixate, speak a word or make a gesture in response to a command. Both PVS and MCS may be permanent or temporary, and the physical examination may not reliably distinguish one from another. Patients typically fluctuate in and out of PVS, the only clue to MCS being a behavior so fleeting that it is easily missed. "By themselves no confirmatory tests provide an accurate diagnosis or prognosis," the AMA asserted.[10] Yet a PVS diagnosis seals one's fate: transfer to a nursing home, where one's basic needs may be attended to (perhaps none too assiduously) and no rehabilitation will be attempted.

There are also scales of functioning, such as the Glasgow Coma Scale, that attempt to quantify the depth of unconsciousness. Though some consider it the gold standard, Ellen Kaufman, MD, of Northampton, Massachusetts (whom we will meet again later in this chapter) calls it "a scaffolding on which to hang our ignorance—not unlike the Greeks naming the gods."

Based on these criteria—essentially, the absence of all cognition—ethicists can invoke the "right to die," as determined by the PVS patient's wishes or the interpretation of them by the patient's surrogate. This was the case with Carrie Coons, an eighty-six-year-old woman from Rensselaer, New York, who was in a vegetative state for five and a half months following a stroke. She had often stated her wish to die rather than be a "vegetable," and, honoring that wish, her sister applied to a judge for permission to withdraw her feeding tube. Permission was granted, but on April 9, 1989, before the order could be implemented, Mrs. Coons woke up, started speaking, and said she didn't wish to die after all. The Supreme Court of New York State vacated the order. She ultimately died in November 1991, never having recovered sufficiently to leave the nursing home.

The vegetative state is more difficult to determine than brain death since one cannot look for flatlined brain waves or the absence of intracranial blood flow, according to Dr. Ford Vox, writing in *Slate* in 2009.[11] Consciousness and its shades of gray can elude confirmatory tests, including standard MRIs, CT scans, and EEGs, he points out. Complicating the picture are cases of brain-injured patients, such as Kate Bainbridge, who have full awareness but whose motor functions are so disabled that they appear unresponsive.

NOIRTIER DE VILLEFORT

Which brings us to locked-in syndrome (LIS). This is a (supposedly) rare condition in which the brain stem, which connects the brain to the spinal cord, is irreparably damaged but the cortex is healthy. The sufferer is entirely lucid but imprisoned within an inert body, unable to speak or move a muscle—except perhaps to blink an eyelid.

Whereas PVS reflects damaged cerebral hemispheres and an intact brain stem, LIS is the opposite—cerebral hemispheres largely intact, parts of the brain stem disabled. The term was coined by Dr. Fred Plum. In his book *Diagnosis of Stupor and Coma,* he describes a patient with arteriosclerosis of the basilar artery who, like the rather sinister, wheelchair-bound M. Noirtier de Villefort in Alexandre Dumas's *The Count of Monte Cristo,* a "corpse with living eyes," is awake but incapable of communicating except with eye blinks. Eerily, Jean-Dominique Bauby had just been thinking about Noirtier de Villefort, toying with the idea of writing a "modern, iconoclastic version" of the novel, when he suffered a devastating stroke in December 1995.

It left him paralyzed from head to toe. He was forty-three years old, at the top of his game as the editor of French *Elle,* and an intellectual and man of culture. Now he could communicate only by blinking his left eye. One blink meant yes, two meant no. He awoke to find himself at the naval hospital at Berck-sur-Mer, on the English Channel. "As soon as my mind was clear of the thick fog with which my stroke had shrouded it, I began to think a lot about Grandpapa Noirtier. I had just reread *The Count of Monte Cristo,* and now here I was back in the heart of the book, and in the worst of circumstances."

He quickly discovered he was utterly helpless, pinned in bed by the heavy "diving bell" of his inert body. Fortunately, his caretakers were alert enough to notice his attempts to communicate by blinking and to recognize that he was in locked-in syndrome (LIS). Recovery, he learned, would be glacially slow, if it happened at all. "If the nervous system makes up its mind to start working again, it does so at the speed of a hair growing from the base of the brain. So it is likely that several years will go by before I can expect to wiggle my toes." Bauby's goals were quickly pared down to the most elementary: to eat and discard the nasogastric tube, to regain the ability to breathe normally without a respirator, to be able to muster the breath to make his vocal chords vibrate so that he could speak. "But for now, I would be the happiest of men if I could just swallow the overflow of saliva that endlessly floods my mouth."

That sentence may have taken half a day to communicate by a series of eye blinks. With the help of a devoted speech therapist, Bauby worked out a code consisting of a letter board with the letters positioned according to their frequency of usage in French. The therapist would point to a letter, and the author would signal yes or no with an eye blink. In this slow and laborious way, Bauby managed to dictate an astonishingly beautiful memoir, *The Diving Bell and the Butterfly.*

He struggled daily with despair. Listening to his children's voices over the telephone, he was anguished to have been struck mute. Being shaved without an adequate amount of shaving cream was a torture, and he was powerless to protest. He tried to keep his mind sharp and avoid descending into resigned indifference, but it was difficult. Physically he was more helpless than a newborn. If the alarm for his feeding tube went off, he'd be stuck next to its infernal beep-beep-beep until someone noticed. Moving about (or, rather, being moved) was cumbersome in the extreme. "After every day's session on the vertical board, a stretcher bearer wheels me from the rehabilitation room and parks me next to my bed, where I wait for the nurse's aides to swing me back between the sheets." The stretcher bearer always wished him *"Bon appetit!"*—which was ironic given that "in the last eight months I have swallowed nothing save a few drops of lemon-flavored water and a half teaspoon of yogurt, which gurgled noisily down my windpipe."

In his mind, however, Bauby was a frequent flier. Leaving

his heavy, inert body, his mind could fly, like a butterfly, anywhere in space and time. He could journey to Tierra del Fuego, visit the woman he loved, build castles in Spain, discover Atlantis, shoot into deepest space. At night his dreams were compelling and complex. "Since you never return to reality," he writes, "your dreams don't have the luxury of evaporating. Instead they pile up, one upon another, to form a long ongoing pageant whose episodes recur with the insistence of a soap opera."

Published two days before the author's death in 1996, his book went on to become a best seller in Europe and later a movie.[12]

How many patients in the United States are locked in?

"Probably thousands," estimates Richard Melia of the National Institute on Disability and Rehabilitation Research, noting that this represents a small fraction of stroke victims.[13] Known causes of LIS are stroke damage to the brain stem, amyotrophic lateral sclerosis ("Lou Gehrig's disease"), and Guillain-Barré syndrome, a self-limiting demyelinating syndrome. Guillain-Barré patients generally recover, ALS patients never do, and LIS patients may show slight improvement but their prognosis is poor.

Dr. Marinos Dalakas, of the NINDS, says that LIS cannot be confused with brain death "because in brain death there is no movement. The patient is not aware at all."[14] There can be confusion with PVS, although PVS patients are incapable of purposeful movements and are unaware of their surroundings.

"The best experts can be confused," he admits. "These are natural lesions and they are not designed to follow our definitions. There are gray areas; sometimes with PVS there is information going through the brain." (An interesting admission, running counter to the party line.) "Most patients in LIS cannot move their eyeballs but can blink. They respond to threatening stimuli, such as when you move your hand close to their eyes. PVS patients do not.

"Part of the brain stem is knocked out, but the part of the brain stem responsible for wakefulness is intact."

Could there be a person with cognitive functions intact who cannot blink his eyes? It was an obvious question, but it seemed to rub Dalakas the wrong way. "It's a very hypothetical situation if a person was not able to blink but can still think," he snapped. "Are you writing a work of fiction?"

Meanwhile, in an Indian hospital a fourteen-year-old boy with transient LIS due to Guillain-Barré syndrome narrowly missed being pronounced brain dead.

MISDIAGNOSIS

Upon admission to the hospital, the boy rapidly slipped into a coma. He was placed on a ventilator, and his condition deteriorated to the point where his pupils were fixed and dilated and all brain-stem reflexes and deep tendon reflexes were absent. "The patient met the criteria for brain stem death," the authors observed. ". . . We performed the apnoea and cold caloric tests supporting the diagnosis of brain death." An EEG, however, showed some sign of brain activity. By the eleventh day the patient's pupils began to react and he began to improve gradually. Five months later, he was breathing on his own and had full recovery of speech, though he was still wheelchair-bound due to severe muscle weakness. He had had the most severe case of Guillain-Barré syndrome on record.

"Brain death is the current medical definition of death when the other body organs continue to function and is defined as an irreversible cessation of all functions of the entire brain including the brain stem," the authors write. In this case, however, "an EEG supported the presence of cerebral cortical function."[15] Note that EEGs are not required to declare a patient brain dead. Are we seeing a misdiagnosis in the areas of PVS and LIS also?

"Is it possible that PVS is the same thing as locked-in syndrome, but without purposive eye movements?" asks Chris Borthwick, an unaffiliated researcher in Australia, in a paper critiquing the diagnosis of PVS.[16] His suggestion that all, or even most, cases of PVS are really LIS without the eye blink seems far-fetched, but it is reasonable to suppose that *some* so-called vegetative patients may actually be in LIS. If they are cortically blind and/or unable to move an eyelid, they will have no way to signal from inside their prisons. Borthwick is an independent researcher from Australia—not a doctor—whose background is history and law.

One prevailing myth is that PVS can be diagnosed on the basis of a bedside assessment. In 1993 a study examined all patients admitted in the previous five years to the Healthcare Rehabilitation Center in Austin, Texas. These patients were all at least one month postinjury and had been diagnosed as comatose or vegetative. Eighteen patients (37 percent) were found to have been diagnosed inaccurately.[17]

Chillingly, a 1996 study at the Royal Hospital for Neurodisability in London[18] found that of forty patients referred to as being in the vegetative state, seventeen (43 percent) had been misdiagnosed. Seven of the patients had been presumed to be vegetative for longer than one year, including three patients for more than four years. All of the patients remained severely physically disabled, but nearly all were able to communicate their preferences as to quality of life issues, some at a high level. All seventeen patients tested in the "severe" level of the Glasgow outcome scale. For fifteen of them, pushing a buzzer was the only functional movement. They ranged in cognitive ability from five (confused, inappropriate) to eight (nearly normal) on the eight-level Rancho Los Amigos Scale used to evaluate patients emerging from coma. Eleven (65 percent) were able to carry out simple one-stage or two-stage mental arithmetic tasks ("What is ten divided by five?").

"It is disturbing to think that some patients who were aware had for several years been considered to be, and treated as being, vegetative," the authors note.

"The vegetative state," they add, "needs considerable skill to diagnose, requiring assessment over a period of time; diagnosis cannot be made, even by the most experienced clinician, from a bedside assessment. Accurate diagnosis . . . requires the skills of a multidisciplinary team."

The key points, according to the authors, are three:

- Most of the misdiagnosed patients were blind. Since strokes frequently result in blindness, the absence of eye blinks or visual tracking is not a reliable diagnostic sign of PVS.

- Any motor activity, no matter how slight, can be used for communication by the profoundly disabled person.

- Identification of awareness requires a multidisciplinary team experienced in long-term management of disability due to brain damage.

A failure rate of 37 to 43 percent is certainly unacceptable. It calls into question the categories used to separate wheat from chaff on neurology wards and suggests that a multidisciplinary assessment, probably including a PET scan, should be required to diagnose PVS.

In reaction to this study, Ronald Cranford, a member of the original Multi-Society Task Force, pointed out that all seventeen patients found to be aware were severely paralyzed and many were blind. Thus "reasonable people may differ in their views of the quality of life of these conscious individuals, but I would speculate that most people would find this condition far more horrifying than the vegetative state itself, and some might think it an even stronger reason for stopping treatment than complete unconsciousness."[19]

We would all sleep better at night if we could believe that patients in unendurable situations were unaware, but that does not make it so. A 1993 study states, "'The PVS patient may 'react' to painful stimuli, but he or she does not 'feel' pain in the sense of conscious discomfort of the kind that doctors would be obliged to treat or of the type that would or should seriously disturb the family."[20]

Though the brain stem is intact, the authors explain, the cerebral hemispheres have suffered irreversible damage, expunging all consciousness and volition. That would be more convincing if (1) a diagnosis of PVS were absolutely watertight, and (2) if we knew "everything there is to be known about the neurology of consciousness and pain," as Chris Borthwick observes.[21] How can anyone, no matter how expert, "conceivably know that a person with PVS experiences no pain despite evincing pain behavior?" Though Borthwick is not a scientist and some of his theories go too far, he asks reasonable questions.

Meanwhile, in the nursing homes where PVS patients end up warehoused, the staff caring for them is encouraged to believe that they have the IQ of couches and can't feel pain. (Based on *"Cogito, ergo sum,"* René Descartes declared that animals could not feel pain.) According to *The Merck Manual* on vegetative state, "More complex brain stem reflexes, including yawning, chewing, swallowing, and, uncommonly, guttural vocalizations, are also present. Arousal and startle reflexes may be preserved. . . . Eyes may water and produce tears. Patients may appear to smile or frown. Spontaneous roving eye movements . . . may be misinterpreted as volitional tracking and can be misinterpreted by family members as evidence of awareness."

Those misguided family members!

Most popular literature on the subject adopts a patronizing, all-knowing tone. A 2005 pamphlet assures the reader that PVS patients "can appear to be making deliberate gestures or to be at least minimally aware. Thus, it can be very difficult for family members or loved ones—and even, at times, for health care professionals—to understand that the patient is not, in fact, aware."[22]

These assumptions (harking back to the standards adopted in the MSTF consensus statement) put the onus on patients to *prove* that they are aware. The two main questions—Do they feel pain? Can they recover?—are removed from the debate by building irreversibility and lack of consciousness into the very definition of PVS. Pain behavior in a vegetative state is thus disregarded as meaningless.

When does a vegetative state become permanent? According to the MSTF, it is when it has lasted for "several weeks." Later the document states, "A patient in a persistent vegetative state becomes permanently vegetative when the diagnosis of irreversibility can be established with a high degree of medical certainty." A page later, we learn that this essentially boils down to "when a physician tells the patient's family, with a high degree of medical certainty, that there is no further hope for recovery, or, if consciousness were recovered, the patient would be left severely disabled."[23]

In other words, you are in a permanent vegetative state if you will (probably) be either unconscious or disabled.

Mingled with the reasonings of Plum, Jennett, and Cranford is the argument, overt or implied, that these patients, even if not *technically* vegetative, are an encumbrance. They place a great financial burden on society, serve no useful purpose, and have lives we can't bear to contemplate. It would be best for all concerned if they could be put out of their misery. (Whatever happens, the pamphlet cited above assures the family that their loved one won't suffer. "By definition, patients in PVS do not have the capacity to suffer because they lack self-awareness.")

There's another problem: PVS, unlike brain death, is not recognized as death in any legal system. To grant permission to allow the patient to die, courts in the United States and Great Britain require a demonstration that any recovery of cognitive functions above a vegetative state is *impossible.* Relatives can petition the courts for permission to withdraw a feeding tube, as happened all too publicly in the case of Terry Schiavo in Florida. (An autopsy ultimately revealed that her brain matter had been largely destroyed.) An article in *Journal of Medical Ethics* stated that "the obligation to keep a person with PVS alive may be overridden, we believe, when we are morally certain that their condition is irreversible."[24]

But even if we assume that a person in PVS has lost awareness, either PVS is not always irreversible or the diagnosis is subject to error. How can one say with absolute certainty that a patient has no chance of improvement, when some (albeit a tiny minority) improve or even wake up years later?

DISCONNECTION SYNDROME

Ronald Reagan was president when Terry Wallis was last aware of the world. The Berlin Wall was standing, Eastern Europe was Communist; there was no Internet, e-mail, or wireless, and no one had heard of Osama bin Laden.[25] In 1986, Wallis, a seventeen-year-old car mechanic from Kentucky, was in a car accident and suffered severe brain damage. For the next nineteen years he was sunk in a minimally conscious state, very disabled, able to respond only with head nodding and eye contact. One day his mother, visiting him in the nursing home, asked him, as usual, "Who is visiting you today?" She was just making soothing conversation and didn't expect an answer.

"Mum," he said.

Over a three-day period Wallis went from articulation of a single word to near-fluent speech. He is still disabled, has no short-term memory, and needs help with many tasks. Although he remembers his mother, he doesn't remember that he had a wife and finds it difficult to understand that the nineteen-year-old woman who comes to visit is his daughter, who was an infant when he last saw her.[26]

"Determining if someone is conscious is at least as important as determining if they are alive," says neurologist Nicholas D. Schiff, the director of the Laboratory of Cognitive Neuromodulation at Weill Cornell Medical Center in New York City. "When I was in medical school twenty years ago, the one-liner we were taught about brain injury was 'What's gone is gone.'"[27] That is changing.

"Clinical practice shows how puzzling it is to recognize unambiguous signs of conscious perception of the environment and of the self in [patients with severe acute brain injury]," wrote Steven Laureys, a neurologist and expert on coma at the University of Liege, Belgium, in 2002.[28] He was in the first phase of a historic collaboration of British and Belgian researchers to scan the brains of vegetative patients with PET. Because PET works in real time, the researchers were able to map the patients' brain activation in response to auditory signals (clicks) and "noxious stimuli."

Surprisingly, many areas of intact neural functioning remained in most patients. Those islands of normality were not, however, integrated with other islands of intact functioning elsewhere in the brain. Laureys proposed that a vegetative state may be seen as a "disconnection syndrome." Whether these neural patterns correspond to awareness remains an open question.

The collaboration continued. In 2004 Adrian Owen, the man who found Kate Bainbridge with a PET scan, made functional MRI (fMRI) scans of patients with sound stimuli delivered through earphones. An fMRI can reveal blood and metabolic activity in the brain in real time. In this study, several patients presumed to be in MCS demonstrated that they could make sophisticated semantic distinctions. For example, one sentence began, "The shell was . . . ," at which point a person would normally visualize a seashell, but the rest of the sentence made it clear that the shell in question was an artillery shell. The neural patterns of the patients indicated that they recognized the difference, according to Owen.[29]

In a three-year study, Belgian and British neuroscientists scanned twenty-three patients earlier diagnosed as vegetative. One twenty-five-year-old British woman had been in an apparent vegetative state for five months and was totally unresponsive. In the fMRI she was instructed to imagine herself playing tennis. Her premotor cortex lit up, the same area seen in normal volunteers who imagine they were playing tennis. Next she was told to imagine herself wandering through the rooms of her house—and, as in normal volunteers, a spatial region of the brain, the hippocampal gyrus, sprang into action. In short, her patterns of neural activation were identical with those of normal volunteers.[30]

Four of twenty-three patients (in their first group of supposedly vegetative patients) were found to be "aware." But *how* aware? "To be able to do what we have asked, you have got to be able to understand instructions, you have to have a functioning memory to remember what tennis is and you have to have your attention intact," notes Owen. "I can't think what cognitive functions they haven't got and still be able to do this."[31]

By 2009 a twenty-nine-year-old Belgian man was grabbing the headlines. Rom Houben had apparently been dead to the world since a car accident five years earlier, but the fMRI told a different story. First Houben demonstrated the ability to imagine playing tennis and to wander around his house. Once those signals were established, the scientists added a twist. They devised a yes/no code, using simple biographical questions such as "Is your father's name Thomas?" Houben would answer "yes" to this by imagining playing tennis. If the answer was no, he did not imagine playing tennis. He used his brain patterns to answer yes/no questions about his family with 100 percent accuracy, even after the code was switched, making tennis the "no" and navigating through his house the "yes."

Having spent five years as one of the living dead, Houben had a story to tell. "Powerlessness. Total powerlessness. At first I was angry, then I learned to live with it" was his first utterance (via facilitated communication).[32] He had normal brain activity, according to Laureys and his team, and was the superstar of a combined British/Belgian study published in the *New England Journal of Medicine* in 2010. Five of the fifty-four patients (diagnosed as vegetative or minimally conscious) were able to willfully modulate their brain activity, and three of them revealed some form of awareness in a bedside assessment. "These results show that a small proportion of patients in a vegetative or minimally conscious state have brain activation reflecting some awareness and cognition," the authors concluded.[33]

The hope is that such tools will permit communication with patients who can't communicate physically, so that they can be asked the important questions, such as whether they are in pain or even if they want to live or die. The scanner can also reveal why certain patients aren't responding to aural stimuli—if the auditory pathways have been disrupted, for example. The test will show whether they are deaf or blind, even when they can't speak.

Videos appeared of Houben with his eyes closed, typing surprisingly quickly, with a therapist guiding his fingers over the keyboard. This prompted some people to think of a Ouija board and to wonder whether his messages were *too* facilitated. A debate broke out. Steven Laureys said he stands firm in his assessment of Houben's intelligence and felt sorry for his patient. "He has gone from being ignored all these years and considered vegetative . . . and now he is being treated as if 'it is impossible, he cannot be a cognitive being.' "[34]

Critics point out that by raising the specter of awareness, this study might either give false hope or deny closure to relatives who have already mourned their loved ones. This does not alter the fact that something important has occurred. "This changes everything," says Nicholas Schiff. "Knowing that someone could persist in a state like this and not show evidence of the fact that they can answer a yes-no question should be extremely disturbing to our practice."[35]

AMBIEN AWAKENINGS

But can anything be done for such patients beyond recognizing that they are more aware than the furniture around them? George Melendez was spending the night at a motel in New York State with his mother, Pat Flores. George had lingered in a minimally conscious state since he had crashed his car into a pond and almost drowned five years earlier. He was barely responsive. His doctors had told his mother, "This is as good as it gets." Since he was thrashing and moaning that evening she gave him an Ambien so he could sleep. Shortly afterward, she became aware of a silence; she looked over at her son, expecting him to be asleep. Instead his eyes were open and looking around the room. He seemed fully alert. "George?" she said.

"What?" he answered. It was the first time since his injury that he had spoken. His mother started giving him Ambien every day and asked him questions—Did he know where he was? Did he know what had happened to him? Was he in pain?—and he was able to provide clear answers. He told her he was not in pain, which came as a great relief to her.[36] He was brought in for a consultation with Nicholas Schiff, who compared PET scans of Melendez before and after Ambien. "His frontal lobe was now bright red; there was a twofold difference. His brain was turned on with this stuff! I've seen at least three other cases of Ambien awakenings.

"It is crucial to develop a biological model for how consciousness recovers," he adds. He believes the key to awakening lies deep in the brain in the area of the thalamus and neighboring structures such as the corpus striatum. Ambien works on certain receptors to remove inhibition from an area of the brain called the globus pallidus that is in the same vicinity. He and his colleagues are very interested in George Melendez and other real-world Rip Van Winkles.[37]

Another was Don Herbert, a Buffalo, New York, firefighter. While battling a house fire in December 1995, he became trapped by debris and suffered head trauma and near suffocation. After a heartbreakingly brief awakening early in the course of his illness he slipped back into a minimally conscious state for the next nine years. His children could hardly bear to see him in this state, but his wife carted him to major family events, where he sat slumped in his wheelchair, seemingly far removed from the world.

His wife begged the doctors to tell her what his chances of recovery were. "Well, look at him," one said. "What do you see? There's nothing there." Nine years passed and then one day Herbert woke up. The nursing home called his family, and all his family members and his firefighter buddies rushed to his bedside. There was much rejoicing, bittersweet though it must have been. Someone brought a cake. A nurse made a video. Herbert's voice was hoarse and stilted after nine years of disuse, but he can clearly be heard asking, "How long have I been like this?"

"Ten years," he was told. The wonderment and grief on his face spoke volumes. Although blind, Herbert readily recognized people by their voices—all but one person. His youngest son had been four when he had lost consciousness; he was now thirteen, tall and deep-voiced. When Herbert put out his hand to touch the top of his head, he grossly miscalculated. Everyone kept saying "Higher . . . No, higher . . . Higher than that." (Herbert died in 2006.)[38]

Extraordinary as it was, Herbert's recovery was not entirely spontaneous, according to Schiff. His medications included a cocktail of L-dopa drugs, the principal medication for Parkinson's disease, and they probably woke up his brain. The brain damage in Parkinson's is in the substantia nigra, deep in the brain, which makes dopamine and communicates with the nearby thalamus. Without dopamine compounds Parkinson's patients sink into a stupor. In recent years brain pacemakers have been implanted in thousands of Parkinson's patients. Schiff began to wonder if the idea could be extended to traumatic brain injury.

It had to be the right patient—impaired enough to warrant a gamble but alert enough to follow simple instructions and give consent. The perfect patient was a thirty-nine-year-old man who had been living in a Cleveland nursing home since suffering brain trauma in a mugging. Though stabilized, he had not improved in the five years he'd been in the home. He could follow simple instructions but had no speech, was fed by a feeding tube, and spent the day with his eyes closed.

What followed was a collaboration among researchers from the Cleveland Clinic, the JFK Johnson Rehabilitation Institute Center for Head Injuries, and Weill Cornell Medical College. The procedure, called deep brain stimulation (DBS), was performed at the Cleveland Clinic by Dr. Ali Rezai, who had previously implanted brain pacemakers in patients suffering from deep depression, obsessive-compulsive disorder, and Parkinson's (one of his patients was the journalist Michael Kinsley). After drilling small holes in the skull, he threaded the wires into the thalamus, the area presumed to need a wake-up call. Then he placed a pacemaker under the skin of the patient's chest and threaded the wires up to connect with the ones in the brain. When the switch was thrown, the patient got dramatically better.

He was now alert, with open eyes, and he could follow instructions. If asked questions he answered them consistently and reliably. His personality came back, and he reconnected with his mother. He could speak, eat, drink from a cup, watch television without falling asleep. He can say, "I love you, Mommy." He doesn't seem to be laying down new memories, but his old memories are intact. "We don't know yet if it will work in other patients," Schiff acknowledges. "And we don't know for sure why it works."[39]

SOMETHING'S NOT RIGHT

The majority of severely brain-damaged patients, of course, don't "wake up" one day and rejoin the world. My friend Bill Lawren never did, but his story is instructive. (He was mentioned in chapter 1 as the interviewer of Robert Trivers, the neo-Darwinist.)

A fifty-six-year-old writer living in Amherst in western Massachusetts, Bill enjoyed reading the obituaries in the weekly paper because the people there commonly lived into their eighties and nineties. He talked often of having bad genes—the men in his family keeled over from heart attacks at an early age—and he was determined not to waste his life working a regular job. He began writing early in the morning, took a long break for breakfast, then returned to writing until noon. He usually spent his afternoons playing tennis or teaching it. In 1995, he published *The Zone,* a diet book coauthored with Dr. Barry Sears, which has sold more than 2 million copies in hardcover alone.

Bill appeared to be in the pink of health when he went to play tennis with a friend on March 19, 1999. In the middle of the match he stopped and said, "Something's not right," then collapsed on the tennis court from a heart attack. He was resuscitated, but his brain had been deprived of oxygen for eight to ten minutes. Bill's prognosis was bleak, and he made the usual journey from the ICU to the back wards of the hospital to a series of disheartening nursing homes. He gave little sign of being responsive and was written off as vegetative; later his diagnosis was changed to minimally conscious.

Dr. Ellen Kaufman, his significant other, sensed that Bill was in there somewhere. But the head of the ethics committee at the hospital—whom Kaufman came to call "Dr. Death"—predicted he'd last about eight weeks. There was no consensus among the nurses in the ICU on whether "Bill" still resided in the comatose body in the hospital bed. One said no, there was nobody home. Another nurse talked to Bill every day.

When Bill developed pneumonia, he was not given proper X-rays and was discharged to a nursing home after two weeks. "No one paid attention to him there," Kaufman recalled. "He was dehydrated. His tongue had deep furrows and he was not urinating." She got him back to the hospital, where he was intubated immediately, and then she found another nursing home.

This is not a Rip Van Winkle story. Bill never "woke up." He died in a nursing home nearly a year after his collapse. However, as far as Kaufman is concerned, a miracle happened. A friend gave her a copy of a book by Arnold Mindell about communicating with people in coma. For decades Mindell, a radical Jungian based in Zurich, has questioned the belief that "comas are unconscious, inaccessible states and that those in persistent vegetative state are—or should be considered—dead." Those opinions reflect only our limited ability to communicate with the comatose state, he proposes.[40]

"In fact, people in comas resemble mythical heroes. . . ." he writes, who journey "through the outermost gates of reality seeking information in the unknown reaches of existence. . . . Many of these patients venture to the heights and depths to find some degree of ecstasy, prophetic insight, and insight."[41]

Using his intuition, Mindell worked with comatose, dying, and brain-damaged patients to construct a system of communication. It was built on recognizing, amplifying, and interpreting the smallest signals from the patients. A yawn, a cough, a groan, the movement of a finger, a knitting of the eyebrows, a change in breathing rate, spasms, twitches, jerks, any might contain a treasure trove of information.

Through Mindell, Kaufman was put in touch with Gary Reiss, who came to Bill's hospital room and trained her to enter Bill's inner world. She would press his hand in tune with his inhalations and exhalations. She learned how to treat every minimal cue as a communication. When his eyes were moving around, she'd say, "Oh Bill, I see that too!" She felt herself drawing closer to him and found that he was more aware than the doctors thought. He'd turn to look at people when they talked to him; he vocalized. When she made a joke about their being a typical Jewish couple—"I'm doing all the talking"—he smiled.

After nearly a year, Kaufman put together a dream team of advisers, with Bill in the room, to gauge his progress and chances of recovery. The meeting was chaired by none other than Dartmouth's James Bernat, perhaps America's leading proponent of brain death. As to whether Bill should be declared a DNR, Bernat thought not. Bill had been improving. "He's a young man. Give him a couple of months," he said. But Bill's ex-wife insisted that he had valued his intellect and would not want to live like this. Kaufman noticed that "Bill turned his head toward her and gave her a baleful stare."

When Bill died, Kaufman, an obstetrician/gynecologist, was aware of a "sacredness" entering the room, like a birth. "I'm no longer afraid to die."

Determining consciousness is an age-old epistemological problem known as "the problem of other minds." Having no direct knowledge of the mental states of others, we infer consciousness from a person's words or other signals, according to John F. Stins and Steven Laureys.[42] But in cases of severe brain damage we can no longer rely on our intuitions concerning consciousness in others. A brain-damaged patient is often incapable of speech or any volitional movement.

In 1950, at the dawn of the field of artificial intelligence (AI), the British mathematician Alan Turing addressed the theoretical problem of intelligent machines and devised an operational measure in the form of a test. If a human is engaged in a question-and-answer session with a computer and cannot decide if he is conversing with a person or a machine, the machine has passed the Turing test. Clinicians who try to determine if a patient is aware by behavioral measures are implicitly employing a Turing test, the researchers say. They are looking for a word, systematic eye blinks—something. In the vast majority of brain-damage cases, consciousness is a black box.

Most patients in PVS today aren't scanned. If they are tested, it is probably with a regular MRI or standard EEG, which usually does not reveal residual brain function. After receiving a bedside diagnosis, they are warehoused in nursing homes. If they flicker into MCS from time to time or if they are really locked in but unable to move their eyelids, this may never come to light. Unless a family member spends enough time by the bedside to recognize the subtle signs—eyes that track for a minute, an attempt at speech, a lifted forefinger in response to a question—they could live out their lives slumped in a wheelchair parked in front of a television that no one is watching.

Plum died in 2010, and his professional partner for more than forty years, Dr. Jerome B. Posner, continued to claim that

PVS patients "look conscious but are unconscious." He defended Plum's PVS work in the face of the stunning triumphs of Owen and others in finding consciousness in PVS patients. Posner said that their concept of the vegetative state was still correct, but "it means that our diagnoses are not as accurate as we used to think they were."

In defining the persistent vegetative state, Plum and Jennett—and the Multi-Society Task Force after them—were ultimately trying to reduce the question of whether to keep a human being alive or unplug him to a scientific formula. With a diagnostic inaccuracy approaching 50 percent, it would seem that they should have paid more attention to the medical facts and left such ethical, legal, and societal questions to those more equipped to handle them.

The netherworlds of coma reveal our profound ignorance about what the mind is and what constitutes consciousness. For Arnold Mindell, what looks like a vegetative state from the outside is experienced on the inside as a journey through other worlds. How much cortex, how many neurons do you need for that? As we will see in the next chapter, the answer is not clear; it may even be possible to have consciousness without a working brain.

SIX

The Near-Death Experience

Where are NDEers getting all that bandwith?

I FELT a little sheepish about going to a near-death experience group without having died. I feared that it would be like trying to infiltrate Sigma Chi without knowing the secret handshake, and maybe I would be discovered and ejected. This group, one of many regional gatherings under the umbrella of the International Association of Near Death Studies (IANDS), meets on the second Tuesday of every month at the University of Connecticut Medical Center in Farmington. After waiting in the third-floor corridor for ten to fifteen minutes for the janitor to come and unlock the door, we were ushered into a room smelling of lemon floor wax and chalk. As it turned out, I needn't have worried about my near-death creds. There were about thirty of us. As we went around the room introducing ourselves it became evident that people who had had NDEs were in the minority.

An elderly man with a professorial look said, "I flatlined three times and don't have any memories of it."

"I had an NDE but I didn't go to the light," said a middle-aged woman wearing a peasant blouse.

A young man with a prominent Adam's apple said, "No NDE but I'm waiting for one."

"I don't think I had an NDE but I had a dream one night that could have been one," said a woman with very black hair who turned out to be a psychic.

Great! I would fit right in.

The NDE has been around a long time. Plato told of a soldier who returned from death with an edifying story of a future life, and in the eighth century the Venerable Bede related the story of a Northumbrian householder who died one night and was revived in the morning, having met a man "of shining countenance and bright apparel" who gave him a tour of hell. The French philosopher Michel de Montaigne experienced the pleasure of dying while he lay benumbed after being thrown from his horse. Dr. David Livingstone had a peaceful and dreamy NDE in the jaws of a lion, and the British philosopher A. J. Ayer had an NDE while in the hospital with pneumonia; he described his experience in the London *Sunday Telegraph* but assured the atheist community that he was still an atheist.

I know what you're thinking: This is where the book veers off into the spiritual, where I offer up possibilities of a hidden realm or Heaven or an ancient Egyptian paradise. Or at least reincarnation, *something* beyond death. Like the Charles Addams cartoon in *The New Yorker* depicting two new angels: one says to the other, "I knew about the wings, but the webbed feet are a surprise."[1] About life after death, I make no hypothesis. As a science writer I've seen no evidence for it or against it. No, the NDE's relevance to death is that it is not death but a sign of life. For many years the very existence of the NDE was considered laughable. People were thought to be making those things up. We are now in an era where even hard-nosed skeptics admit that people who have had close brushes with death sometimes have those visions.

The debate is only over what they mean. Are they evidence of another unknown version of reality? Or can they be explained in mundane physical terms? Neither explanation is safe for medical science. The first explanation takes doctors out of their realm of expertise, where they are helpless and clueless like the rest of us. The second explanation, which many embrace, is even more threatening. It keeps the NDE in physicians' antiseptic world but puts the lie to their definitions of death. Yes, the NDE may turn out to be some kind of hallucination. But dead people don't dream or hallucinate, and many NDE voyageurs have been pronounced dead, either via circulatory criteria (after which the brain is supposed to go to black momentarily) or via brain death, which is supposed to mean that the "whole brain" has shut down. Which begs the question: if the NDE is a hallucination, what is hallucinating?

Perhaps the NDE is generated by the untested part of the body between the brain stem and the toes. Or perhaps the brain operates just fine without a blood supply. Whatever, the NDE tells us not to rely on people in white coats with stethoscopes around their necks to give us the last word about life and death. If the reader believes in an afterlife, I'm not here to argue with him. It's beyond my salary range. It may just include a bright light and lost relatives. For me, I suspect it will mean webbed feet, and I will meet the smug members of the Harvard ad hoc committee. Chairman Beecher will greet me, saying, "Yeah, we were just winging it."

CONSCIOUSNESS AFTER DEATH?

The modern era of near-death studies began in 1975 with Dr. Raymond Moody's popular book *Life After Life,* based on testimonials of 150 people who had been narrowly rescued from death. Moody coined the term "near-death experience" and identified a core narrative, which goes something like this:

A man is dying and as he reaches the point of greatest physical distress, he hears himself pronounced dead by the doctor. He hears an uncomfortable noise, a loud ringing or buzzing, and feels himself moving rapidly through a long, dark tunnel. After this, he suddenly finds himself outside his physical body, and he sees his own body from a distance. . . . He watches the resuscitation attempt as if he were a spectator. He begins to collect himself and notices he has a "body" but one of a different nature. At this point other beings come to meet him and he glimpses the spirit forms of dead relatives and friends. A warm, loving being of light surrounds him and he is shown a panoramic replay of the major events of his life. At some point he approaches a border or barrier and is told that he must return to his body. He resists at first, enveloped as he is in profound feelings of joy, love, and peace, but does reunite with his body.

Later he tries to tell others about his experience but can find no vocabulary adequate to describe his unearthly experiences. Others scoff at him, so he stops talking about it. Still, the experience affects him profoundly.[2]

Inspired by Moody's book and anxious to bring scientific rigor to the subject, Kenneth Ring, a psychologist at the University of Connecticut, called hospitals in the Hartford area looking for people who had "died" and come back. He studied a hundred patients who had had experiences similar to those described by Moody: a sensation of great peace, calm, and joy, the cessation of pain, perception of the body from above, floating through a tunnel, going toward the light, a life review, meetings with dead relatives or friends and with divine beings, coming to a border or barrier and being turned back.[3] His study confirmed Moody's findings and showed that NDEs do not favor any particular race, gender, age, education, marital status, or social class. Religious affiliation has no bearing on whether a person has one.[4]

"There is no question that the NDE exists," says Ring. "People who are utterly sane do undergo a compelling experience. Even skeptics concede that. The debate is about what it means. I don't think these experiences in any way prove an afterlife. However, on the basis of my studies, I'm prepared to conclude I don't think consciousness ceases when our body dies."[5]

Ring subdivided the NDE into five stages, maintaining that 60 percent of people experience stage one (peace and contentment) while only 10 percent experience stage five (entering the light). He also developed a Weighted Core Index, while Bruce Greyson, a psychiatrist with the University of Virginia Health System, devised a Near-Death Experience Scale, which helps differentiate between NDEs and non-NDEs and rates different depths of experience.[6]

By definition, though, death is irreversible, which means that if you come back you were never dead. Although delving deeply into this borderland, most researchers stop short of claiming that the NDE is a snippet of "life after death." Our focus in this chapter is a deceptively simple one: is the so-called NDE evidence that consciousness can exist apart from a working brain? If so, if our neurological machinery is unnecessary to mentation, we may have to admit we don't know what death is, or personhood, and we should certainly reexamine our standards of brain death.

Let's return now to the IANDS meeting in Farmington. After the introductions, a guest speaker was announced. She was a lively middle-aged woman named Sarah who had had an NDE more than a decade before. She had been driving to work on I-95 in the morning rush hour, and the last thing she recalled was an SUV rolling over and heading her way. She learned later that she had been thrown from her car and landed on her head. Passing motorists who saw a woman flying through the air thought they were seeing the filming of an action movie. A state trooper came to her aid. He performed CPR and she began to breathe again.

That was Sarah's first clinical death. The second time her heart stopped and was restarted occurred soon afterward in the medical helicopter (when she recalled being "with a divine being and looking down at the Earth and feeling very loved and safe"). She underwent a third clinical death in the hospital's trauma unit. Each of her "deaths" was medically documented (allegedly), and each generated a different segment of her complex NDE.

Unlike many NDEers', the memories of her NDEs came back to her gradually, weeks and months after she emerged from her coma. "I went through a tunnel, but it was huge. Lining the sides of the tunnel I saw every person and every animal I had ever loved. The light was there. There was a curved bridge—like one of Monet's bridges—and on the other side I saw my mother, wearing the apron she wore on her baking days. My father was there, younger and more handsome than he had been in life. 'Go back! You have to go back!' my mother yelled in the same screechy voice she would use when she would call us in to supper. 'Honey, don't you know? It's not your time!'

"Now, do my parents really look like that? Of course not. They've discarded their bodies; they don't need them anymore. You are shown things in a way that you can understand."

In the hospital trauma unit where Sarah's heart stopped for the third time, she found herself floating near the ceiling looking down at twenty people working feverishly on her body, trying to save her life. "I looked down and it was just the body of a woman. I didn't know it was me. You come back into the same body you had before, and it's really weird. I was asking questions as best I could. 'Who are you? What is this? Can I get out of here?'

"I was in a coma for a while. You can hear when you're in a coma. I heard the doctors talking. They were saying, 'She isn't going to function at a very high level.' I was pissed. I thought, 'I'll show you SOBs.'

"When I woke up I said to the doctor, 'I heard what you said.' I'd been intubated so I sounded like a cross between Donald Duck and the Godfather. He apologized. I said, 'No, I want to thank you. You really motivated me.' "

A 1982 Gallup Poll found that 8 million Americans (4 percent of the population) had had an NDE and now there is no shortage of support networks for those who have returned from the dead. If you log on to the IANDS website, for example, you can see a large collection of vivid NDE accounts. In contrast, the IANDS definition of the NDE reads like a nonprofit corporation's mission statement, vague and globally inclusive. For example, it accepts "NDEs" that are unconnected with death. (It's not my business, but doesn't that dilute the concept?) Below are portions of three representative NDEs culled from the archives:

> The pain suddenly stopped and I just popped out of my body and floated up to the ceiling. I could see the

dust on top of the light fixtures, and I thought, "Boy someone's going to catch it for this." I could see doctors working on someone on the table, when all of a sudden I realized it was me—my body. My family was down the hall and I wished my children would stop crying. I wanted to let them know I was fine, but they couldn't hear me. [A woman who had a heart attack in a hospital emergency room]

A call to 911 rushed me to the hospital. Sometime after a nine-hour operation I had an NDE. I entered a tunnel and had a life review. I then found myself being escorted by this loving Being of Light to the most beautiful garden imaginable. Describing it is hard . . . how can we describe beauty that comes from within? It has been almost three years now and it is as vivid to me today as it was in the very beginning. [A woman who suffered a ruptured aneurysm]

When I entered into the tunnel there was no attachment to nor memory of the physical world as we know it. . . . It was an experience of total acceptance. . . . Another unforgettable thing was the lack of time and space. . . . I was at one and the same time in the "past" (medieval times), "present," and "what will be." It was as if the three time frames were superimposed [like] numerous layers of film laid one on the other. . . . I tried to insist on staying, saying I absolutely did not want to go back. [I was told that] "Gerry must go back because they don't know." I reentered my body through the top of the head. [A woman struck by a car][7]

The Dutch cardiologist Pim van Lommel is the author of an exhaustive study of 334 patients who suffered cardiac arrest and were resuscitated. Of them 44 (18 percent) had a near-death experience. "When the heart stops beating, blood flow stops within a second," he reports.

> Six-point-five seconds later EEG activity begins to change due to shortage of oxygen. After fifteen seconds there is a straight flatline and the electrical activity in the cortex has disappeared completely. We cannot measure the brain stem, but animal studies also show that the activity stops there.
>
> Moreover, you can prove that the brain is no longer functioning because it regulates reflexes, such as pupil response and the swallowing reflex, which no longer respond. So you can easily slip a tube down someone's throat. The respiratory centre also shuts down. If the individual is not reanimated within five to ten minutes, their brain cells are irreversibly damaged.[8]

Van Lommel's study was published in the British medical journal *Lancet,* a periodical not known for its paranormal bias. But based on his research, Van Lommel has come to the conclusion that the brain neither produces consciousness nor stores memories. Think of a television set: when it is broken or damaged, programs may no longer come through it, but the source is not inside the set. In the same way, consciousness may not be a product of the brain.

Dr. Peter Fenwick, a neuropsychiatrist and NDE expert at the University of Southampton in the United Kingdom, is drawn to the same conclusion. "The brain isn't functioning. It's destroyed. If you faint, you don't know what's happening. The memory systems are particularly sensitive to unconsciousness, so you won't remember anything. Yet [after an NDE] you come out with clear, lucid memories. This is a real puzzle for science."[9]

EXPLANATIONS

But isn't there a perfectly good scientific explanation for all the "dreamlike" reports of return from death? Skeptics point to the fact that NDEers were dying and their brains were presumably malfunctioning from lack of oxygen, drugs such as morphine, or seizurelike activity. The visionary events, they say, are most likely a last fireworks display of a dying brain—just before the final curtain drops. Or the mechanism may be psychological: confronted with death, the patient is terribly stressed and simply hallucinates a more pleasant reality. Part of that hallucination takes the form of depersonalization—a separation from one's body. In either case, it is a hallucination and in no way real.

A prominent skeptic, Susan Blackmore, a psychologist at the University of Plymouth, in the UK, is convinced that NDEs are caused by a combination of physiological and psychological reactions, triggered by disturbed brain function and/or stress.[10] The tunnel is caused by oxygen starvation, she maintains, which kills certain brain cells. Because people cannot explain away NDEs, they turn them into an afterlife. To accept NDEs as paranormal experiences, or glimpses into another world, flies in the face of physics, biology, and logic, according to Blackmore. "I suggest that there is no stream of vivid pictures that appear in consciousness. There is no movie-in-the-brain. There is no stream of vision. And if we think there is we are victims of the grand illusion."[11]

Leaving aside other objections, if Blackmore's psychological theory were correct, you'd expect religious believers to have a higher incidence of NDEs, but they are no more apt to have them than are atheists and agnostics, people who apparently have no need to believe in an afterlife.

Skeptics are apt to airily dismiss NDEs as a result of a lack of oxygen to the brain or the many drugs that dying patients may be given. But the scientific evidence does not bear this out. Anoxia is a particularly unlikely explanation, as we'll see.

Here is a list of the major mechanisms that have been proposed to explain NDEs:

1. Lack of oxygen (hypoxia) in the brain or excessive carbon dioxide (hypercarbia)

2. Temporal lobe seizures

3. Endorphins, the natural opiates responsible for "runner's high," or similar drugs.

4. Ketamine, a potent dissociative anesthetic (and sometime recreational drug) that is known to produce out-of-body experiences.

5. Depersonalization, in which people replace the unpleasant reality of approaching death with pleasant fantasies

6. Birth memory, a stored memory of birth replayed at death

7. REM intrusion, also known as dreaming while awake; a recent article in the journal *Neurology* noted that some characteristics of REM sleep (rapid eye movements, low muscle tone, and dreaming) can occur during wakefulness and proposed that

people who have NDEs are predisposed by vulnerable brain arousal systems[12]

But each of these explanations becomes weaker when we scrutinize it:

1. **Lack of oxygen.** NDEs have occurred when monitors show the brain to be oxygen-saturated. Oxygen deprivation is associated with chaos, fear, and confusion and does not jibe with patients' correct and clear accounts of what happened around them while they were "dead." Nor is anoxia associated with bliss.

2. **Temporal lobe seizures.** These have been known to trigger hallucinations with some NDE-like features, but temporal lobe epilepsy is known to produce states of fear, sadness, and loneliness. The supposed out-of-body episodes produced by temporal lobe stimulation, when examined closely, are unlike NDEs. (We'll return to this.)

3. **Endorphins.** These suppress pain and cause a high, but they do not precipitate hallucinations. Morphine and other opiates do, but they lack the lucidity of the typical NDE report. Some studies show that patients who are on narcotics or Valium are less likely to have an experience.

4. **Ketamine.** Although ketamine is famous for producing out-of-body hallucinations, many ketamine trips are marked by fear, confusion, and alienation. (We will come back to ketamine later.)

5. **Depersonalization.** The depersonalization theory, first proposed in 1930, is an old standby. The basic idea is that dying

people freak out and dissociate from their bodies. However, the peacefulness and clarity of most NDEs argue against it.

6. Birth memory. A baby being born does not float toward a light at great speed. Birth is a long, probably painful progress propelled by contractions. The baby's head is not positioned so that it can see a light ahead, and it is hard to credit the doctor as being the prototype of the "being of light" reported during NDEs.

7. REM intrusion. Research has shown clear differences between NDEs and hallucinations. People who dream while awake recognize immediately that the dreams were not real, whereas NDEs are commonly described as "realer than real." In the study cited in *Neurology,* 40 percent of people who had NDEs did not report ever having had an episode of REM intrusion.[13]

"None of these reductionist explanations is compelling, though some are plausible," Greyson told me. "I don't have much use for people who just spew theories."[14]

Van Lommel notes, "Up until now, 'death' simply meant the end of consciousness, of identity, of life. . . . In the past, these experiences were attributed to physiological, psychological, pharmacological or religious reasons. So to a shortage of oxygen, the release of endorphins, receptor blockages, fear of death, hallucinations, religious expectations or a combination of all these factors. But our research indicates that none of these factors determine whether or not someone has a near-death experience."[15]

All of the "explanations" posit that the NDE is some sort of hallucination. But hallucinations tend to be illogical, bizarre, fleeting, and distorted, and NDEs are usually orderly, lucid, logical, and comprehensible. NDEs lead to profound personality transformations that last for decades. Furthermore, according to NDE researcher Melvin Morse, who studied NDEs in children, "Nothing explains the light."

The most arresting characteristic of the near-death experience is its vividness and clarity; the NDEer is certain it was not a dream. When you wake up from a dream, you know you were dreaming and that you are now awake. You may not be able to prove it but you don't doubt that you are awake. Let's take a closer look at two purported causes of NDEs, temporal lobe seizures and ketamine.

SEIZURES AND "VITAMIN K"

In 2002 four Swiss physicians published an article in *Nature* claiming that "the part of the brain that induces an out of body experience (OBE) has been located." The pseudo OBE occurred during focal electrical stimulation of the brain of a forty-three-year-old female epileptic, who was awake during the procedure (the brain itself does not feel pain) so that she could report her sensations when an electrode was applied to different areas. This is known as "brain mapping." Stimulation of the right angular gyrus near the right temporal lobe provoked a series of odd bodily perceptions. "I see myself lying in bed but I see only my legs and lower trunk," the patient said when the doctors increased the electricity to the area. More electricity, and she saw her legs becoming shorter. "When asked to look at her outstretched arms during the electrical stimulation . . . the patient felt as though her left arm was shortened; the right arm was unaffected. If both arms were in the same position but bent by 90 degrees at the elbow, she felt that her left lower arm and hand were moving toward her face. . . . When her eyes were shut, she felt that her upper body was moving toward her legs, which were stable."

The authors describe OBEs as "curious, usually brief sensations in which a person's consciousness seems to become detached from the body and take up a remote viewing position." This rather bare-bones definition allows them to claim, "These observations indicate that OBEs . . . can be artificially induced by electrical stimulation of the cortex," and to speculate on how this might occur during a near-death episode.[16]

But do these perceptual distortions really resemble a spontaneous OBE? Researchers Janice M. Holden, Jeffrey Long, MD, and Jason MacLurg, MD, argue that they do not.[17] Compare the following spontaneous out-of-body experience by an English patient anesthetized for foot surgery:

> Before coming round I saw myself up in a corner of the room and I was looking down upon the hospital bed. The bedclothes were heaped up over a cradle and my legs were exposed from the knee down.
>
> Around the right ankle was a ring of plaster and below the knee was a similar ring. These two rings were joined by a plaster strip [on] either side of [the] leg. I was struck by the pink of my skin against the white plaster.
>
> When I regained consciousness two nurses were standing at the foot of the bed looking at the operation, one quite young. They at once left the private ward and I managed to raise myself up and look over the cradle, seeing again exactly what I had seen when still "out."
>
> . . . The particular way in which the plaster had been applied was plainly seen from my position in the corner of the room and the contrast between the white plaster and the pink skin was striking.[18]

Obviously, the Swiss case pales in comparison with the complex experience of a spontaneous OBE or NDE. Furthermore, bodily distortion is not a feature of spontaneous OBEs/NDEs. Reviewing hundreds of reports submitted to various NDE websites, Holden et al. found no distortions of body image or illusions of bodily movement. Spontaneous OBEs are also distinguished by lucidity and are deemed by the experiencer to have been real. (Presumably, the epileptic woman in the first case recognized her experience as unreal.)

Since stimulation of the temporal lobe mimics a few aspects of the NDE, such as leaving one's body behind or having memories flash before one's eyes, some researchers have hypothesized that the stress of hurtling toward death precipitates bursts of activity in the temporal lobe. Arguing against this is the fact that temporal lobe hallucinations are generally pervaded by fear, sadness, and loneliness, whereas most NDEs are calm, peaceful, and ecstatic.

Now let's return to the ketamine theory. Some twenty-five years ago the UCLA psychologist Ronald Siegel reported that he was able to reproduce NDEs in his laboratory by giving LSD and other drugs, including the dissociative anesthetic ketamine, to volunteers. (As a street drug, ketamine is known as K or Vitamin K). A skeptic of the NDE, Siegel attributes all its features categorically to a ketaminelike chemical produced by the dying brain. The NDE, like drug hallucinations, he says, is based on the release of stored images. The fact that NDEs have certain core features is explained by the common architecture of the human brain. Thus chemicals catalyze an archetypal experience.[19]

Kenneth Ring says, "Any adequate neurological explanation would have to be capable of showing how the entire complex of phenomena associated with the core experience (the out-of-body state, paranormal knowledge, the tunnel, the golden light, the voice of a presence, the appearance of deceased relatives, beautiful vistas and so forth) would be expected to occur as a consequence of specific neurological events triggered by the approach of death."[20]

At first glance Karl Jansen appears to have come close. A native New Zealander who is a psychiatrist at the Maudsley Hospital in London, Jansen gave ketamine to volunteers, many of whom went on to describe "becoming a disembodied mind or soul, dying and going to another world." Childhood events may also be relived. Users often are not sure they have not actually left. One man had had a near-death experience when he passed out in a burning house and was revived. Later, in grief over his partner's death in the fire, he took ketamine several times "and the same thing happened every time. You could fly and you could actually travel. You are in the place where everybody is who ever died."[21] Jansen hypothesized that K exerts its effects by altering receptors in the brain for the neurotransmitter glutamate. On ketamine, he notes, 30 percent of his subjects believed that their experience of leaving the body was real.[22]

Since his first articles and books, however, Jansen's views have become less reductive. He now believes that both the near-death experience and the ketamine trip are portals to a transcendent reality—to a real God, in effect. "I am no longer as opposed to spiritual explanations of these phenomena as this article would seem to suggest."[23]

Dr. Peter Fenwick of the University of Southampton in the United Kingdom says, "The difficulty with these theories is that when you create these wonderful states by taking drugs, you're conscious. In the near-death experience, you are unconscious."[24]

GOING VERIDICAL

But the best case for the reality of NDEs is the "veridical account" by a patient who, while unconscious, sees precisely where the nurse put his dentures, hears the medical team talking, or sees the medical procedures performed and the instruments used in his resuscitation. These reports can be compared to medical records and the testimony of the medical team. There have been a number of veridical reports over the years, but the queen of all veridical NDErs is Pam Reynolds.

Her name is a pseudonym. ("I'm as famous as I ever want to be," she explained.) A composer and children's book author from Georgia, she was diagnosed with a giant aneurysm of the basilar artery below the brain stem in 1991, when she was thirty-five. Ballooning out like a bubble in a defective tire, the aneurysm, if ruptured, would immediately be fatal. Its size and location put it out of range of conventional surgery. Pam was referred to Dr. Robert Spetzler, the director of the Barrow Neurological Institute in Phoenix, Arizona. Spetzler had pioneered a drastic surgical method called hypothermic cardiac arrest. Surgeons call it the "standstill"; some patients call it the "freeze."

The day Pam "died" was August 6, 1991. Her body temperature was lowered to 60° F by running her blood through a cooling machine. "At around 28 Celsius [82° F] the heart begins to fail," Dr. Spetzler explains. "The bypass machine takes over. At 16 Celsius [61° F] the body is cold enough to survive without any blood flow at all up to sixty minutes. The cooling alone would arrest the heart but we use potassium chloride—the same chemical used in lethal injection—to complete the cardiac arrest, because otherwise fibrillations would damage the heart."

During the hour-and-twenty-five-minute procedure, Pam's body was wired up like a NASA flight control center. Her blood pressure was monitored with a two-inch plastic tube at her wrist; a three-foot-long Ganz catheter was threaded through into the artery to her lung to measure pulmonary pressures and blood flow from the heart. Her heartbeat, as well as her oxygen saturation, was continuously monitored. Her urinary temperature was monitored in her bladder, and her core body temperature measured deep in her esophagus. Her brain temperature was registered by a thin wire embedded in its surface. Standard EEG electrodes recorded the brain waves of the cortex, while small molded speakers placed in her ears emitted a series of clicks to test the electrical activity of the brain stem.

As her core body temperature dipped lower, the doctors watched all signs of consciousness peter out. Her brain waves—both cortical EEGs and those of the brain stem—flatlined, and there followed forty-five minutes when there was no electrical activity in her brain. While other members of the surgical team were executing the various steps of a cardiopulmonary bypass, Dr. Spetzler drilled through Pam's cranium with a Midas Rex bone saw. He exposed the aneuryism and tilted her head up so that all the blood drained out of it. The aneurysm deflated; he clipped it off, then reinfused the brain with blood. For four or five minutes there was no blood in Pam's brain.[25]

Meanwhile, Pam was launched out of her body by the high-pitched whine of the bone saw drilling through her skull. "It was awful," she recalls.[26] "It reminded me of the dentist. Everything inside me was shaking, from my belly button to the top of my head. As a musician I hear my world so I am very disturbed by noise." The next thing she knew, she had "popped out"—a description that recurs in NDE accounts—and was up above her body looking down at it. "I looked at my body and I knew it was mine, but I didn't relate to it being me. It was like a car. It was like, Okay, the car's a wreck but I have insurance—and you just walk away. Every moment of my life I'd spent dragging around that heavy thing that was lying on the table."

She saw the surgeon sawing through her skull with an instrument she would describe as resembling an electric toothbrush with a groove in it and retractable blades. She described the heart-lung machine and heard a female doctor say that her arteries and veins were very small. Her veridical out-of-body perception would turn out to be accurate at a time when every monitor attached to her body showed her to be lifeless.

After that Pam felt herself being pulled through a tornado-like vortex—"It was not a tunnel but it was like a tunnel"—and she was spinning around in it. "It felt like going up in an elevator very fast." She came to a place of light. "It was incredibly bright, like sitting in the middle of a lightbulb." She was surrounded by beings of light that gradually took form as individuals, including her grandmother and several other deceased relatives. "They were feeding me. They were not doing it through my mouth, like with food, but they were nourishing me with something. Sparkles is the best way to describe it. I had the sensation of being fed, nurtured, made strong." At some point she was given a vast amount of information about the future of Earth, which she keeps to herself. ("You don't want to know.")

The beings would not allow her to go farther, communicating nonverbally that if she went all the way into the light "something would happen to me physically. I wanted to go into the light but I also had children to raise." Finally her uncle led her back to her body lying on the operating table. "It looked like a train wreck. I said, 'I'm not getting in that thing.' My uncle had to convince me by talking about all my favorite foods. When I came back, it was like jumping into a pool of ice water. It hurt! At that moment I heard 'Hotel California' playing [about checking out of the hotel anytime but you can't leave]. I mentioned later to one of the doctors that the choice of that song was incredibly insensitive and he laughed."

Could it have been a dream? "This was no dream. It was absolutely real, with smells, colors, sounds. Everything was recorded as clearly as everyday life. More clearly. I was saturated with it. I have never heard or seen anything as vivid as when I was dead. My senses were enhanced; my ability to comprehend was enhanced."

When she returned home, an Atlanta neurologist brought her to the attention of cardiologist Michael Sabom, the author of a landmark two-year study of 160 patients who had been resuscitated after cardiac arrest. ("Forty-seven of them had NDEs of a depth of 7 or better on the Grayson scale.") He recalls, "When I first heard her tell what happened I didn't believe she was telling me the straight story. I had never heard of the procedure. The medical aspects of it alone are phenomenal—draining all the blood out of someone's head and refrigerating the person to sixty degrees and then warming them up, and they'll still be alive?

"In cases where the patient with a flat EEG has an NDE, the skeptics will say, 'Well, she still had blood perfusion in her head.' But Pam had all the blood drained out of her head. She is like a laboratory experiment. She's hooked up to everything; everything in her body is being monitored to the limits of our scientific ability. She was a flatliner in several respects—in the cortex, the brain stem, and heart. During part of the surgery, you couldn't have said if she was alive or dead. The surgeon could have been operating on a corpse at that point; he had no way of knowing. Only later, when she was revived, could you say that she had been alive at the time."

He points out that Pam's brain waves were monitored throughout the procedure and there was no seizure activity. "At least in this one instance we can say for sure that the NDE was not the result of a temporal lobe seizure.

"When NDEers accurately report medical facts while they are 'out,' skeptics say, 'Well, they just overheard conversations subliminally under anesthesia and reconstructed them later,' " he adds. "But Pam had these molded plastic things stuck in her ears to measure auditory evoked potentials from the brain stem—so she could not have heard anything while in her body."[27]

As Dr. Jeffrey Long, a Gallup, New Mexico, anesthesiologist and NDE researcher, puts it, "It is very helpful to personally have an NDE or NDE-like experience. . . . For virtually all NDEers, an NDE cures NDE disbelief."[28]

People who have NDEs almost invariably attest that it was not a hallucination or a dream. They say they *know* it was real and don't care what anyone else says. A typical statement (in this case from a relative of my wife's who suffered respiratory arrest during surgery) is that "it wasn't a dream. It felt real. Everything was so vivid. I could smell the forest I was moving through. In dreams you can't smell."

The real-life feel of NDEs is puzzling. Daniel C. Dennett, the favorite philosopher of neo-Darwinists, explains that video games are not as realistic as the real-life events they represent—football games, car racing, sex, grand theft auto—because they don't have the bandwidth, the daunting amount of information, that reality requires. According to Dennett, "Television requires a greater bandwidth than radio, and high-definition television has a still greater bandwidth. High-definition smello-feelo television would have a still greater bandwidth, and *interactive* smello-feelo television would have an astronomical bandwidth."[29] This, he explained, is why hallucinations and dreams are only a wan replica of reality. No matter how exciting or frightening a dream may be, when we wake up, we know it wasn't real. But those with NDEs say their experiences were real. Where are they getting all that bandwidth?

Studies show that nearly all NDEers lose any fear of death and at least 98 percent believe there is life after death. Common after effects include spiritual growth, a loving attitude, knowing God or a higher power, inner peace, and a sense of purpose in life. More than 80 percent say that their concern for others has increased. Between 59 and 89 percent report having psychic phenomena or healing abilities after an NDE. Some people claim to be able to see into the future as we know from the aforementioned Pam Reynolds. "There are people who see the future during a near-death experience," says Pim van Lommel. "For example, there was a man who saw his future family. Years later, he found himself in a situation he had already seen during his near-death experience. I suspect this is also the way déjà vu works."[30]

Sometimes integrating an NDE into your life can be challenging. Sixty-five percent of NDEers' marriages end in divorce, compared with 40 to 50 percent of the general population. If they are unable to communicate their experiences, these returnees, especially those who have undergone a harrowing experience, may lapse into alienation and depression.

But the majority of the changes are positive and long-lasting. "I used to be a very controlling person," says Sarah, from the NDE meeting. "The NDE made me more accepting. I live my life with purpose. When I go it's going to be so great to get back to that place."

"One obstacle to understanding this subject is a belief that death is an event, a moment in time," says Sabom. "There is no definable moment of death but only a process of dying that starts with life and eventually ends in death. There is a road from life to death, and the NDE is on the road. If you were driving from Houston to LA and stopped over in El Paso, the NDE is El Paso."[31]

Sabom relates the following account by a patient named Brent during cardiac arrest:

> Suddenly I see people running into my cubicle and I'm saying "Not me! Not me!" And obviously they can't hear a thing, but my mind is working. There is all sorts of calamity going on, and I'm telling them, "Leave me alone. Leave me alone!" I thought they should be looking after someone else. . . .
>
> And I'm laying there watching them from where my head was on the pillow and I'm trying to yell at them, "What are you doing? What are you doing?" . . . They were pulling my clothes off and I wanted to hang on to my clothes. I was trying to pull my pants on and someone was ripping them off. . . . And I asked them later if they didn't hear me and they said, "No, you were dead."
>
> But I said, "No I wasn't. My mind wasn't dead. I could see you. I knew what you were doing and when you were coming in I even felt you push the board under me."[32]

"Medical studies have shown that after the onset of a cardiac arrest such as Brent's, consciousness is not lost for an average of 9 to 21 seconds," says Sabom, noting that the shock and pain of resuscitation procedures could literally scare a patient to death.[33]

"Death is not a specific moment," says Sam Parnia, a fellow in pulmonary and critical care medicine at Weill Cornell Medical Center. "It is a process that begins when the heart stops beating, and the lungs stop working and the brain ceases functioning—a medical condition termed cardiac arrest. During a cardiac arrest, all three criteria of death are present.

"Subsequently," he adds, "there is a period of time, lasting from a few seconds to an hour or more, when emergency medical efforts may succeed in restarting the heart and reversing the dying process. The big question is, Can human consciousness continue when we've reached the point of death and all studies have shown that the brain stops functioning? We can't explain how people have consciousness when the brain is flatlined."[34]

Dr. Parnia currently divides his time between New York and hospitals in the United Kingdom, as he leads an innovative scientific study, AWARE (AWAreness during REsuscitation), aimed at studying the brain and consciousness during cardiac arrest to discover what happens when we die. The largest study ever of near-death experiences, it involves twenty-five participating hospitals in the United Kingdom and continental Europe. Perhaps at the end of this study we will have more information about NDEs, but do not expect proof of life after death.

"The afterlife," says Carol Zaleski, a religion scholar at Smith College, "will ever remain beyond the reach of our instruments. All we have is stories."[35] As to the hereafter, there are a number of possibilities, including that some people are immortal while others are not. (That is what the ancient Egyptians and the Gnostics believed, Zaleski says.) Another possibility is that the "afterlife" experience is very brief, the length of a typical NDE, and after that—lights out.

Some Christian fundamentalists are hostile to the standard NDE, which appears equally available to the religious and irreligious and doesn't require anyone to go to Hell or to "earn" Heaven. (No significant correlation has been found between religious beliefs and the likelihood or depth of NDEs.) "On the Internet there are diatribes against Moody and me," Ring says.

But though the first accounts of NDEs were uniformly positive, it now appears that approximately 17 or 18 percent are harrowing, marked by distress, loneliness, confusion, and so on. "Evangelicals have seized on this with glee," Ring says.[36]

Let me also pile on a little here by saying that the 4 percent of Americans figure seems high—that's a lot of people who have been declared clinically dead and then seen the dark tunnel and all that. Some people need attention, need to feel special. I attended another group of NDEers at which an elderly woman talked about how she had had a head-on collision with another car and been lifted out of the driver's seat into the air, where she had witnessed the crash. Then she had been set down unharmed by the side of the road. Another lady said, "That's not an NDE, dear. That's an OBE, an out-of-body experience." I did not make any friends that day when I commented, "It sounds more like a DUI."

NDE research suffers from relying on anecdotal reports. It seems safe to say, however, that as things stand, the NDE has not been explained away by science. But even if the NDE is merely a very deep and convincing hallucination, and if the patient met all the criteria of death, how, then, was he or she hallucinating? If his brain was down—no electrical activity, no blood flow in the case of Pam Reynolds—then *what* was hallucinating? Clearly, there is more going on in the brain than the flash-splash-gasp of the Harvard criteria can begin to examine. People in some cultures, such as rural Greece, consider death to be final after the person has been buried for a year. The NDE raises the possibility that, from the inside, death unfolds over time, from stage to stage.

"I think death is going to be really neat," says Sabom. "I think the NDE is a real experience, but my belief is we're dealing with a spiritual experience that cannot be proved."

SEVEN

Postmodern Death

Death is becoming divorced from biology as many segments of society—the law, the military, hospitals—turn it into a construct for their convenience.

AFTER twenty-year-old Juan Mancia was gravely injured in a gas pipe explosion at an Arlington, Virginia, construction site, his heart stopped while he was being medevaced to Washington Hospital Center, where he was pronounced dead on arrival at 11:36 in the morning. At that point he became a potential organ donor.[1]

He had never signed a donor card, and his family was opposed to donation, but his wife was not at the hospital when he died. Taking advantage of a bizarre D.C. law, doctors performed a "preharvest." To preserve his organs while they awaited formal consent, they made two deep cuts in his abdomen and a larger one in his groin, then spent an hour doing surgery on his body and flushing ice-cold preservatives through his kidneys in the hope his family would say yes. Mancia's wife said no.

Besides having the misfortune to die in D.C., Mancia was a new category of potential donor—a "non-heart-beating cadaver." The University of Pittsburgh started the trend in 1993 by retrieving organs from patients who had not been declared brain dead but were terminally ill and dependent on a ventilator. In the best-case scenario, the death is planned, and the patient or his family gives consent. The patient is then wheeled to an operating room, where his ventilator is removed while the transplant team stands by and waits for his heart to stop beating. (It sometimes doesn't stop, and the patient must be wheeled back to his room.) Once the patient is pulseless, the transplant team counts out two minutes (sometimes more) and declares him dead. It sounds like a short window, and it is.

Other centers followed Pittsburgh's lead, and soon more states began considering legislation like the District of Columbia's, legalizing invasive organ preservation in the absence of family consent. The large pool of non-heart-beating donors could increase organ donation by 25 percent, proponents point out. Critics object that such protocols give a "perverse incentive" to doctors to minimize the care given to terminally ill patients, as their organs become more important than their lives.[2]

With cardiopulmonary death, transplantation must take place quickly, before "warm ischemia" ruins the organs. This may mean prepping the patient for harvest *before* death. Such medical interventions range from giving the patient organ-preserving drugs to inserting cannulas, or tubes, into the arms and groin for the delivery of cooling liquid.

Advance consent is required for the insertion of cannulas

prior to death. However, physicians at the Regional Organ Bank of Illinois, after being refused permission in thirty-five cases, undertook "preservative infusion" (the infusion of cold preservative solution into the abdominal aorta and peritoneal cavity to preserve the organs) without family consent, reasoning that it was nondeforming and nonmutilating and did not require consent. Only then did they approach the families about organ donation, and six of seven consented.[3]

Proponents of cold-preservative "preharvesting," such as was performed on Juan Mancia, say it expands the window of opportunity by several hours so as to locate next of kin and allows them "time to grieve" (and, just incidentally, to donate the loved one's organs).

"I am very nervous about anyone on a mission," says Dan Teres of Baystate Medical Center. "There is a serious potential for a slippery slope with non-heart-beating donors. I would not want the organ bank expert in charge of a patient. Say the family wants to donate but the patient does not cooperate and maintains a low blood pressure. The organ bank expert might consider medication to speed up the process."

In this chapter we'll take a look at what I call, for lack of a better term, postmodern death. I don't really know what "postmodern" means, but I believe it has something to do with the rejection of objective reality. A scientific definition of death is difficult to attain when so many segments of modern society treat death as a construct, defining it in any fashion they find convenient, with no regard for reality. The law, medical examiners, the military, executioners, and others divide the living and dead into two categories using a logic that defies biology. Unfortunately, one group that also ignores biology is doctors. We shall begin with them.

ECMO AND OTHER LOOPHOLES

Forget about the vicissitudes of brain death. Plain-vanilla death, circulatory death, can also be ambiguous. "You can think the heart has stopped, and it hasn't," says one ethicist.[4] After a person has drawn his last breath, he is not necessarily dead right away. Some patients who appear gray and lifeless, with prolonged lack of heartbeat, could potentially be brought back to life. Modern resuscitative techniques have pushed the horizon of irreversibility—the sine qua non of death—further and further back.

Clinical death always *precedes* biological death. A medical examiner observes, "Clinical death usually occurs at the bedside. When the heart stops beating and the person ceases to breathe and when at the same time the pupils remain fixed and dilated and there is absence of tendon reflexes, the physician is legally justified in signing the death warrant."[5] But that is not yet irreversible biological death. From a medical examiner's perspective, furthermore, days or even weeks after the last breath and the last pulse, a kind of life lingers in the body. (We'll come back to this.)

In a perfect world, the withdrawal of life support would always be planned, consent obtained, and all the forms signed. But what if a person arrives DOA in the emergency room, fails to be resuscitated, and is without a donor card or a next of kin? Having no heartbeat or pulse, the patient is in no position to give consent, yet if his organs are to be donated, procedures must be initiated immediately to preserve them.

One such procedure is extracorporeal membrane oxygenation (ECMO). An ECMO machine, similar to a heart-lung machine, continuously pumps blood from the patient through a membrane oxygenator that imitates the gas exchange process of the lungs, removing carbon dioxide and adding oxygen. Oxygenated blood is then returned to the patient.

What if in response to ECMO, the donor, declared dead, begins to show signs of life during the organ procurement? It is not out of the question. A recent study reports that "the use of ECMO in cardiac patients unresponsive to cardiopulmonary resuscitation can result in a 31.6% survival to hospital discharge. Declaring patients dead before initiating ECMO . . . can deprive some patients of the chance of survival and full recovery."[6]

The authors caution, "The probability of return of signs of life or the need for suppression of signs of life during surgical procurement would make the act of procurement an act of homicide or mercy killing." Although malpractice is a constant concern of American physicians, wrongly declaring a patient dead is hardly a worry. That is because the vagueness of the UDDA gives doctors lots of wiggle room. Even so, there are legal dangers, say the bioethicist Joseph Verheijde of Arizona State University and his coauthors, because the courts have yet to render an opinion on "whether persons can be held liable for injuries arising from the determination of death itself." Because the legal and scientific criteria for death are not uniform, it might conceivably be concluded that "organ procurement itself is the proximate cause of death."[7]

At a meeting of intensivists for our benefit at the Baystate ICU,[8] the doctors were discussing non-heart-beating cadavers.

"You get the patient declared, and then they try to resuscitate him and take the organs. It doesn't make sense to resuscitate them first," said one doctor.

"What's the difference?" another doctor shot back. "You're letting a patient arrest and then taking the organs, or you're taking the organs and letting the patient arrest. It's a semantic thing."

"How long do you wait?" we asked.

"You resuscitate them until you get a line in."

"You take the patient off his ventilator, wait for his heart to stop, and then resuscitate him?"

"Some don't arrest, and you have to take them back to the ICU or back to a floor."

I sensed that some of the doctors were confused.

A nurse mentioned that the anesthesiologists tended to be "emotional" about the situation, perhaps because of "inadequate education" about this relatively new procedure. "In one situation the anesthesia staff left the room," he told us.

"In this situation we have to do everything we have been taught not to do," said a doctor. "We stop everything until the patient dies."

BENDING THE DEAD-DONOR RULE

Peggy Rickard Bishop Bargholt became a passionate advocate of organ donation after the organs of her three-year-old son Jeffrey, who died of a brain hemorrhage, brought life to six different children. She began working with Lifebanc, an organ procurement agency in Cleveland, where she made a troubling discovery: the Cleveland Clinic had designed a new protocol to secure organs from non-heart-beating cadavers.

She was interviewed by Mike Wallace on *60 Minutes* in a 1997 show with the scary title "Are Surgeons Taking Organs from People Who Are Not Quite Dead?"

She discovered next that the protocols would allow organ-preserving drugs to be given to patients expected to die after life support is withdrawn. The hospital planned to give living patients massive doses of a drug called Regitine to make their organs healthier for transplant. But this drug is known to block the body's release of noradrenaline, which is crucial to patients fighting for their lives.

Disturbed, she took her concerns to Professor Mary Ellen Waithe, director of the bioethics program at Cleveland State University. "We investigated for months," Waithe told *60 Minutes,* "and the response from everyone I consulted with was 'What are they trying to do? Kill them?'" When she heard that, she knew she had to contact the law.

"'As good as dead' is not dead," says Waithe. "The definition of death is not up to doctors. It is up to the legislature of Ohio."

Accordingly Waithe met with the assistant county prosecutor Carmen Marino, who investigated and had conversations with the Cleveland Clinic. Afterward, the clinic announced that it would not implement the protocol after all. At the same time, it issued the statement: "Regitine is not harmful to the patients" and that their protocol "met the highest legal and ethical standard."

Elsewhere the protocol of taking organs from non-heart-beating donors *was* implemented—at University of Wisconsin Hospital, for example. Dr. Hans Sollinger, then the chairman of the hospital's transplant unit and president of the American Society of Transplant Surgeons, assured *60 Minutes* that the patients were brain dead by the time they came under the knife, and he was confident that they felt no pain. The protocol itself seems a little less sure. "If there is doubt about the patient's ability to feel pain, it is appropriate to prescribe analgesic or pain medication prior to transfer to the operating room," Marino told

60 Minutes. "The protocol we looked at said a morphine drip could be maintained for comfort."

"For whose comfort? The dead person's comfort?" Mike Wallace asked.

"Apparently," said Marino. "That's who's getting it."[9]

SEEKING IRREVERSIBILITY

"The ability to diagnose death unerringly is important for obvious reasons; therefore the criteria for death must pass tight scrutiny," stated Michael DeVita, of the University of Pittsburgh, in a thoughtful paper.[10]

With cardiopulmonary death, he writes, "The question of when death occurs turns as much on how long the heart can stop and still restart as [on] whether it has stopped. If the heart can be restarted, the person is not dead, even while the heart is not beating. The individual might be dead but is not necessarily dead." Thus the ultimate criterion is irreversibility.

If there are no resuscitative efforts, the heart is dead once enough time has elapsed so there is no chance the heart will restart on its own. (DeVita goes through the medical evidence on spontaneously restarting hearts and concludes that two minutes is too short an interval to be sure.) The second kind of irreversibility—"irreversible cessation of function even if resuscitation is attempted"—is trickier and would require a longer death watch.

DeVita observes that our ability to resuscitate the heart is so advanced that the organ can be restarted even outside the body. The first heart transplant by Christiaan Barnard in 1968 demonstrated that a heart can be restarted successfully if stopped and removed from the body.

That does not make the original owner of the heart alive. But it does "demonstrate the length of time that must elapse before one is justified in saying a person's heart could not resume functioning even if resuscitation were attempted," according to DeVita. (Recall that Barnard made the declaration of death himself. Today the transplant surgeon may not declare the donor dead, for obvious reasons.)

In clinical studies, the likelihood of resuscitation in uncontrolled situations is much lower. The chance of success is almost nil if resuscitative attempts follow a delay of ten minutes or longer. DeVita observed, "If the diagnosis of death requires that the heart cannot have the capability of resuming function . . . then death certification should never occur until hours have passed." He observes that this definition is clearly incompatible with the current practice of medicine and our social concept of death.

WHEN IS A PERSON DEAD TO THE MEDICAL EXAMINER?

Since the days of Karen Ann Quinlan it has become easier to pull the catastrophically ill or injured off life support. A dramatic case in point was the 1997 murder of Matthew Eappen, the eight-and-a-half-month-old son of the Newton, Massachusetts, physicians Deborah and Sunil Eappen. Their British au pair, Louise Woodward, was tried and convicted on the assumption that she had shaken the baby, causing his death.

"He died of blunt trauma, a skull fracture," says Michael Baden, the former chief medical examiner for New York City, and host of the HBO series *Autopsy.* "He was not a shaken baby. But the medical examiner's findings were not accepted. Normally, the parents would be on trial. There was a fractured wrist that had gone untreated. Both parents were doctors!" (The judge might have agreed because Louise Woodward, although convicted, was set free for time served.)

"Homicides are often mistaken for SIDS when the family pediatrician gets involved," Baden notes.

Another concern of Baden's is drowned infants. "A parent throws a baby into the water. Was the baby dead when it went into the water, or did it drown?" Baden looks for diatoms, plankton, which get into the body when the baby inhales water. He is compassionate toward mothers who kill their babies. "In France and England they are aware of postpartum psychosis. Here we give them the death sentence."

When is a person dead? I ask him.

"You are dead when death is pronounced officially, by a doctor, or police, or even an EMT. The 'time of death' is when a body surfaces from the lake even though it may have been dead for months. Sometimes the body may have been dead for ten years." You are dead, then, when someone notices? "Yes. It's thought that more people die in hospitals in the morning. That's just when the nurses make their rounds."

In some jurisdictions, he says, postmortems are performed by coroners, who are elected and may be funeral directors; elsewhere, they are done by medical examiners, appointed physicians. It isn't always so easy to determine whether a person is dead or not. Mistakes are made. "A woman sits up in the morgue and says, 'What am I doing here?'"

In the O. J. Simpson case, he says, the medical examiner was used simply as a body removal service. The most important evidence in the case were the drops of blood on Nicole Simpson's back, which he detected in the crime scene photos. "If it's O.J.'s blood, it's the end of the case. If it's A. C. Cowlings', that's interesting. It could not have been Ron Goldman's blood. Nicole was bleeding forward onto the steps. The blood had to be dripped from above. You can tell from the velocity; you get big circles of blood from a cut finger." But the evidence was not collected and tested, so we'll never know.

What happens after death? I ask Baden. "The bacteria in the intestines start leaving the intestines and entering the rest of the body, creating gases. Antibodies break down. There is decomposition. It is a war between bacteria and the tissues drying out. In the desert the body mummifies. In New England, the body turns green and bloated and starts sloughing off skin. Embalming kills the bacteria and slows decomposition."

Baden's favorite play is *Hamlet.* He likes the idea that maggots eat the king's body and are then eaten by a fish, which is ultimately consumed by a beggar. He notes some misconceptions. It is not true that hair and nails continue to grow after death. The skin around the scalp and nails dries out, making the hair and nails look longer.

In his work, Baden says, brain death is irrelevant. "If you have a heartbeat you have life."[11]

TIME OF DEATH: NOT AN EASY CALL

A corpse languishes in a pool of blood near the cabana. The medical examiner (a shapely African-American woman in an evening dress who has just been wrenched away from a black-tie event) bends over the body. She sticks a probe into the liver to determine algor mortis, the cooling of the body after death. She examines the underside of the body for evidence of livor mortis, the pooling of blood that occurs after death. She tests for muscle stiffening, rigor mortis. The stomach contents will be examined in the autopsy. She stands up and pronounces the time of death—between nine fifteen and ten in the evening—

and the detectives look for suspects without an alibi for those forty-five minutes.

It happens that way only on television, according to Jessica Snyder Sachs, the author of the highly readable and interesting *Corpse.*[12] The belief that body cooling is a reliable postmortem timepiece harks back to the nineteenth century. In 1887, a celebrated London pathologist, Frederick Womack, attached to a cadaver's belly a thermometer he claimed was exact to one fortieth of a degree (that would strain credulity even now). He reported a steady temperature drop and devised a mathematical formula that he claimed could fix the time of death within minutes for any corpse. Doctors following his lead but using a simpler math than Womack's established the traditional formula: adding one hour since death for every 1.5° F drop in temperature.

How convenient if it were so, but even Dr. Harry Rainy, a Glasgow physician who was the first to apply, in the 1860s, Newton's law to body cooling, acknowledged that the margin of error is wide—about four hours. A corpse found at 98° F could be anywhere between four seconds and four hours dead.

Dr. Rainy also noted a temperature spike right after death, when Newtonian mechanics would predict the most rapid heat loss. Something mysterious was going on, and eventually it became recognized that when a person dies with an underlying infection—even something that would escape detection in life—the result can be a dramatic temperature spike. Stress (such as might occur when trying to fight off an attacker) will provoke a big spike, too. The complex chemical processes undergone by cells after death can unpredictably add heat to the picture.

Even if pathologists ignore the spikes, they run into the further uncertainties of calculating "how an irregular shaped, irregularly composed object lost heat," says Sachs.[13] Heat loss can vary widely depending on where the reading was taken. A breeze or humidity affects the cooling of the body, as do the surface the corpse is lying on and the fabric of the clothing it is wearing, not to mention drugs, blows to head, blood loss, and other factors.

But what about livor mortis? The pinkish color that marks the pooling of blood on the body's underside appears between thirty minutes and three hours after death. The window for lividity's "fixation"—when the skin no longer blanches if you pinch it—lies between four and twenty-four hours. But the degree of coloration is a subjective call. Skin livor is very hard to recognize on any skin darker than pale Caucasian skin. Anemia or blood loss complicates the picture.

Fortunately, we still have rigor mortis, postmortem muscle lock. In 1929 the phenomenon was traced to adenosine triphosphate (ATP) in the cells. The chemical energy stored in ATP's atomic bonds power not only muscular contraction but virtually every biological function on Earth. However, rigor can never be charted with accuracy. The ultimate arbiter, the amount of ATP present in each muscle group at death, has been found to be affected by the fight-or-flight response, the deceased's aerobic conditioning and a host of other factors.

The O. J. Simpson trial offered protracted testimony over the partially dissolved rigatoni tubes in Nicole Simpson's stomach. But stomach contents were dismissed long ago as the least reliable of all postmortem time scales, according to Sachs.

For a time, considerable hope was attached to the eye's vitreous humor as a marker of time of death. Compared to the rapid chemical changes of other dying fluids, eye jelly changes slowly as the cells of the retina dissolve. In 1963 the Little Rock, Arkansas, medical examiner, William Sturner, published his findings that the values of potassium in the eye fluid of a hundred corpses examined rose in a trajectory that could be quantified simply.

But eye jelly fared no better than other putative postmortem clocks. Over the next two decades, numerous studies were done. "Most found even greater variation in postmortem vitreous potassium levels. Some even showed significant differences between the two eyes of the same corpse," Sachs notes.[14] For the time being, Sachs concludes, we are stuck with a certain uncertainty in fixing the exact time of death.

Even the idea of an exact moment of death may be at odds with reality. A new definition began to emerge in the 1950s of death as a progressive and drawn-out breakdown in the body's oxygen cycle. "There are outposts where clusters of cells yet shine, besieged little lights in the advancing darkness," as the surgeon-poet Richard Selzer put it in 1974.[15]

It seems that the more closely we look at death, the more mysterious it gets. Buddhists, who believe that moving the corpse before three days pass will disturb subtle death processes that are still occurring, may be onto something after all.

CAN YOU MURDER A PERSON WHO IS ALREADY DEAD?

This is a legal debate that dates back 1,500 years; it was discussed in the Talmud. If a person is thrown out of a fifth-story window and someone stabs him as he passes the third floor, did he murder the guy? I decided to ask the celebrated Harvard lawyer Alan Dershowitz, who was involved in the first U.S. case of a man accused of murdering a corpse. "Say you hit or shoot a guy who goes into a coma and later dies," he said. "When is it murder? It depends on the state you live in. If it's Maine, it's a year and a day. If your victim lives for a year and a day, you're off the hook for murder."[16]

In 1977 Dershowitz successfully defended a man who had been convicted of shooting a corpse.[17] On a Friday night in 1973 Melvin Dlugash and his friends Mike Geller and Joe Bush went out drinking and partying. Joe had been staying with Mike in his basement apartment in the Flatbush section of Brooklyn. During the evening Mike asked Joe several times to pay his share of the rent, as he had promised. Joe became truculent. The argument escalated after the three men returned to the apartment. At 3 a.m., after several hours of drinking, Joe brandished a long-barreled .38-caliber revolver and aimed it point-blank at Mike. Mike backed away and Joe shot him three times in the chest. Mike fell to the floor, blood gushing from his wounds.

Joe then turned to Mel and said, "If you don't shoot him, I'll shoot you." After some hesitation, Mel went over and fired a bullet from his own gun, a .22, into Mike's head. The next morning, Mel, who was an epileptic, had just had a seizure on his sister's couch when a homicide detective appeared and questioned him. He explained that when he had shot Mike, Mike was already dead. He was arrested and charged with murder.

At Mel's trial the principal medical witness for the prosecution was asked the following question: "Assuming the deceased were shot [in the chest] with a .38-caliber weapon [Joe's gun]. And further assume that five to eight minutes later the deceased was [shot in the head] with a smaller-caliber weapon [Mel's gun]. Doctor, with any degree of medical certainty could you tell this jury whether the deceased was alive at the time that the latter shots were fired into his head?"

The doctor went silent for a moment. Then he said, "No. I couldn't answer that."

The other expert witness, the chief medical examiner for the

City of New York, had overseen the autopsy. He testified that the .38 bullet had torn a path through Mike's lung and into his heart, crossing it from left to right. Asked if that bullet would have been fatal, he said yes.

The trial judge asked if that meant death was instantaneous. The medical examiner said, "No, not necessarily." Asked whether he could state with any degree of medical certainty that the victim would have been alive three to five minutes later, he said no.

The defense ultimately decided not to let the defendant testify, which meant that there was no record of his saying that he had believed he was shooting a dead man. His lawyers also declined to put Joe Bush on the stand, feeling he would be a wild card. The case went to the jury. The decision rested on "a most profound issue of morality and science: when does life end?" says Dershowitz. Having instructed the jury that it could convict Mel of murder only if it found that Mike was alive when Mel shot him, the judge explained that if the jury did not so find, it *could* find Mel guilty of attempted murder. The jury's verdict was unanimous: guilty of attempted murder.

In an appeal, Dershowitz contended that it was impossible for a jury to conclude beyond a reasonable doubt that Mike had been alive when Mel shot him. Further, the brief argued, it could not be concluded beyond a reasonable doubt that Mel had *believed that* Mike was alive. If he had not believed that Mike was alive, he could not have intended to kill him. A third defense hung on the fact that when the assistant DA had interviewed Mel, he had cut him off when he was explaining what had happened. An affidavit had been prepared of what Mel would have said had he not been cut off.

The oral argument was argued before the Brooklyn Appellate Division. On March 1, 1976, the court concluded that the People had "failed to prove beyond a reasonable doubt that Geller had been alive at the time he had been shot by the defendant." It ruled, further, that Mel could not be convicted of attempted murder, since he had not intended to kill a corpse.

THE HOSPITAL OFF THE HOOK

The aforementioned Matthew Eappen was not brain dead, yet the hospital pulled the plug on the orders of the parents, both respected physicians, making examination of the victim difficult for the defense. Could the defense in the Louise Woodward case have moved to dismiss the murder charge since the child's death had resulted from the hospital's or parents' actions, not the defendant's?

"No," says defense attorney and Innocence Project founder Barry Scheck, "if the defendant did something to contribute to the death."[18]

In another case, he tells me, a man was shot nonfatally but malpractice in the hospital resulted in his death. "The court ruled that it was reasonable to assume malpractice would occur, and thus the assailant was guilty of murder." The act of sending a man to a hospital was in effect murder.

HOW TO SURVIVE YOURSELF

Mistakenly declared dead following a cruiser accident in 1982, Patrolman Charles E. Peck, of Springfield, Massachusetts, sued to be named his own survivor and collect death benefits (instead of merely a disability pension). A number of state politicians took Peck's side, but the mayor blocked the measure. "I'd venture to say he's probably the only man in the state who wanted to be treated as if he wasn't there," quipped Mayor Michael J. Albano.[19]

BATTLEFIELD TRIAGE

The point of triage is to "conserve the fighting strength of the troops," says anesthesiologist John Neeld, who was an army medic in Vietnam in 1967–1968 and was wounded in the foot ten days into the TET offensive. (He can still feel the shrapnel under the skin.)

In a strict sense, triage sorts out the wounded in a hierarchy of survivability coupled with urgency. The badly wounded but salvageable (e.g., suffering a mortar round to the abdomen that punched a hole in the intestine) get medical help first. The wounded who can wait and still survive are in the middle. At bottom are the worst cases, those whose future is dim. For all practical purposes, the military considers them dead.

In the ER a new doctor will see a patient from a car accident who is cold with no heartbeat. The doctor will be afraid of making a premature move so he'll needlessly order an EKG. "In the field," says Neeld, "you have to determine that the guy is going to die. I'm going to move on."

Since the Revolutionary War, the U.S. military has employed these pragmatic standards, which have little to do with whether a person is alive or dead.[20]

LIFE AND DEATH IN THE ICU

While in the ICU my friend Bill Lawren developed untreated pneumonia and painful bedsores and was shockingly dehydrated. He was not brain dead or dying; he was breathing without a ventilator and gave signs of recognizing his loved ones. There was still a chance that he could recover to some degree and certainly no reason to suppose he did not experience pain. But his case was considered unpromising; his brain had been deprived of oxygen for seven minutes and the ICU staff had written him off. Like many other ICU patients, he was simply being warehoused until death came.

Having seen lots of doctor shows, we think of the ICU as a place of advanced technology and driven personnel devoted to saving life at whatever cost, where four or five frenzied doctors crowd around a patient yelling "Stat!" and shocking him or her back to life. For many patients, however, the ICU is a medical backwater. For the first few days there may be a flurry of tests and lifesaving measures, but that changes when the patient is judged hopeless. Then no one bothers. And then the main function of the ICU may be to warehouse potential organ donors.

The statistics tell the tale. One study of two ICUs attests to a dramatic change over a five-year period. In 1988, 51 percent of ICU patients died because medical treatment was withheld. By 1993, that number had shot up to 90 percent.

"Only a short time ago, brain death was considered to be the sole acceptable basis for withdrawal of 'life-sustaining' treatment; that is, only the occurrence of death itself justified cessation of efforts to prevent or postpone death," according to Judith E. Nelson, MD, and Diane E. Meier, MD. Today, they claim (citing a study by T. J. Prendergast et al.) that "not only do physicians now find it acceptable to withhold or withdraw life-sustaining therapies in certain cases, but they assert it is inappropriate for families to demand treatment that physicians themselves had considered morally imperative only 15 years before."[21]

Why the shift in attitude? The authors believe that once

brain death became equated with death per se, hospital personnel felt free to withdraw life support earlier, with few qualms and fewer guidelines as to when. In the process they also grew more intolerant of families demanding treatment for patients they have written off. And as ICU doctors come to see their patients as on the road to death they prescribe fewer painkilling drugs, and that may lead to further disease and complications. The judgment "as good as dead" becomes a self-fulfilling prophecy.

The standard of brain death has clearly had a cascade effect on patients who are far from brain dead. On paper, the authors note, the policy on what justifies the withholding of life support is fairly uniform. In practice, it is all over the map. In four hospitals in Minnesota and Wisconsin observed during the late 1980s and early 1990s, 84 percent of patients died as a result of a decision to forgo some form of potentially lifesaving treatment. The process was gradual and incremental; on average, at least three or four interventions were withheld. Decisions to forgo cardiopulmonary resuscitation typically came earliest, followed by decisions about dialysis and vasopressors; limitations of feeding tubes and mechanical ventilators typically came last. Doctors in teaching hospitals were quicker to withdraw or withhold life-supportive measures than those in community hospitals—either because their patients were sicker or because there was a weaker bond between the physician and the family.

Many patients have a living will that includes a "do not resuscitate" (DNR) order. But what does that mean? Traditionally a DNR refers only to cardiopulmonary resuscitation and does not necessarily apply to other critical care interventions (e.g., dialysis, antibiotics, ventilators, blood transfusions, intravenous fluids). Yet in practice the line is blurred. Much depends on the physician's attitude.

A prior relationship with the physician appeared to make

withdrawal of life support less likely. Although private patients were more likely to have a DNR in the record, patients without a private physician were more apt to have life-sustaining measures withdrawn. In patients who had ventilators withdrawn, there was great variability in the administration of morphine and sedatives. "In the ICU as elsewhere suffering may remain untreated because of unwarranted concern about addiction," the authors write. "Suboptimal symptom management not only compromises the comfort of critically ill patients but may also impede recovery, since physiological stress responses are maladaptive and further complicate the course of critical illness."[22]

Many physicians still regard death as failure and approach issues of dying with fear, guilt, anxiety, and avoidance. A "good death," say Nelson and Meier, must be acknowledged as an appropriate goal of medicine; the dignified and gentle death of a patient can be seen as a medical accomplishment of considerable merit.

DON'T LOOK FOR ORGANS HERE

Despite the legality of brain death, says Dr. Robert Kennedy of Maimonides Medical Center, "a death certificate is only issued here on cardiopulmonary death." In any case, the average family at his hospital is hard to convince of brain death. Maimonides sits smack in the middle of Borough Park, Brooklyn, and its two dominant patient groups are Orthodox Jews and Chinese Americans.

"We treat ultrareligious people," says Kennedy. "Orthodox Jews believe all the stops should be pulled out to keep someone alive." The fact that a Jewish funeral must be held within twenty-four hours of death does not leave much of a window to discuss the subtleties of brain death. "We are not a good market for organ donations," he concedes. "To discuss death is to wish death on them."

In a Chinese family the eldest son makes the medical decisions. "We can't talk to the older person. It would be demeaning and insulting. The patient will say, 'Talk to my family—my son.'

"Death is now a legal definition, not a physiological one," says Kennedy. "When you bring in the issue of futility it just confuses the situation."[23]

NOT JUST FALLING ASLEEP

Legal executions demonstrate, first of all, how difficult it is to kill a human being. The electric chair was first used in the 1890s; it was not a pretty picture then and is scarcely better now. In the 1983 execution of Alabama convict John Evans, after the first jolt of electricity, sparks and flames erupted from the electrode attached to his leg. The electrode then burst from the strap holding it in place and caught on fire—and so on for fourteen minutes, grisly detail after grisly detail, until Evans was dead. There are numerous other such electrocutions on the record.

In 1898 Dr. Joseph Alan O'Neill wrote an article, "Who's the Executioner?," describing the electrocution of Martin Thorn in Sing Sing. When Thorn arrived on the autopsy table after three rounds of electrocution, Dr. O'Neill was not certain he was dead. He concluded as follows, "The law requires the post-mortem mutilation. It is, in fact, a part of the penalty; for, as it reveals no cause of death and teaches nothing of interest to science, it is evident that its purpose is to complete the killing." O'Neill claimed that most electrocuted prisoners actually died from the autopsy.

In a modern electrocution, the electrodes are attached to

the prisoner's head. This is meant to be humane, but evidence suggests that the skull insulates the brain so the initial current doesn't cause instant unconsciousness. Boiling of body fluids and burning cause extreme pain if the person is conscious. Mutilation is not uncommon—chunks of flesh, bones exposed—all of which, opponents say, clearly violates the Eighth Amendment prohibiting "cruel and unusual punishment."

Gas is no improvement. The prisoner is restrained in a chair inside an airtight chamber. The executioner opens a valve that allows hydrochloric acid to flow into a pan behind the chair. He then adds potassium cyanide or sodium cyanide crystals into the acid, producing puffy white clouds of lethal hydrocyanic gas. In 1983, in a Mississippi gas chamber, officials had to clear the witness room eight minutes after the gas was released when convict Jimmy Lee Gray's desperate gasps repulsed the witnesses. He died banging his head against a steel pole in the gas chamber, now out of sight of the witnesses.[24]

In Delaware and Washington State, a condemned person has a choice between lethal injection and hanging. If the latter is chosen, the condemned person is weighed prior to execution, as a specific amount of force must be applied to the neck in relation to the person's weight. If this is done properly, death occurs by dislocation of the third or fourth cervical vertebra. The noose is placed behind the prisoner's left ear so as to snap the neck upon dropping. If done improperly, the person may strangle slowly; if he is dropped too far, he will be decapitated.[25]

In Utah, the one state that offers death by firing squad, child killer John Albert Taylor, wearing a black hood, was strapped into a steel chair twenty-three feet from six executioners. A target was placed over his heart and a pile of sandbags behind him, and six men fired at his heart. Death by firing squad may be the quickest way to be executed, but it is evidently not instantaneous even with six bullets to the chest.

Our confused attitude toward legally sanctioned killing is reflected in the fact that if the prisoner suffers a heart attack en route to the execution chamber, he will be revived and then killed. The law states that a person must be sane to be executed, but the Supreme Court ruled that a murderer could be drugged to make him sane enough for execution.

All states that have capital punishment now offer lethal injection. Since the first lethal injection in 1982, more than a thousand prisoners have been executed by this method, according to Amnesty International USA. It looks as if the person goes to his death as tranquilly as if he were falling asleep, but frequently this is an illusion.

Lethal injection is actually a series of three injections. The first drug is sodium thiopental, a short-acting anesthetic, which puts the person to sleep for four or five minutes. The second drug is pancuronium bromide, which paralyzes the skeletal muscles but does not affect the brain or nerves. By paralyzing the diaphragm, it also causes suffocation. The third drug, potassium chloride, the chemical in road salt, travels to the heart and stops it.

If everything is textbook perfect, the prisoner will be unconscious by the time drugs two and three hit his system and death will be painless. However, many real-world variables can alter the outcome, beginning with the fact that lethal injection tends to be carried out by people who are not doctors or nurses. (Medical personnel are barred by the Hippocratic oath from participating in killing someone, though sources intimate that this rule is sometimes broken.)

Whatever is actually happening is masked, however. If the

convict is conscious no one will know because the second drug, pancuronium bromide, paralyzes him and he is unable to speak, move, or cry out. No one understands this better than Dr. Edward Brunner, an anesthesiologist at Northwestern University Medical School.

In a brave act of self-experimentation, Brunner had a physician friend inject him with pancuronium bromide, placing a bag of oxygen over his nose and mouth so that he would not actually suffocate. Even so, it was horrible. "I know what the drugs do. I know what it feels like to be suffocated. It felt like a horse was sitting on my chest. You need movement of the chest wall to satisfy the sensation of breathing." He thought he was screaming to his friend, but he wasn't. He was paralyzed.

"The executioners don't know how to use these drugs," he says. Pancuronium bromide acts like the arrow poison used by Amazonian Indians. A person injected with it remains conscious but cannot speak or move. His diaphragm paralyzed, he suffocates. "It is for public relations," says Brunner, "to make the execution look nicer. The patient would otherwise be thrashing around."[26]

If the anesthetic wears off or doesn't take at all, the prisoner will feel exactly what Brunner did, with one additional feature: there will be an excruciating surge of potassium burning its way through the prisoner's veins toward the heart. The chemical destroys tissue if it is accidentally spilled. Hospital patients who have been injected (accidentally) with potassium chloride scream in pain before dying.

"It would basically deliver the maximum amount of pain the veins can deliver, which is a lot," says Columbia University anesthesiologist Mark J. S. Heath. And all the while the person may appear to be asleep.[27] (Not always, however: In 1989 a Texas convict, Stephen McCoy, died gasping, choking, and heaving—a reaction so violent that it caused one of the witnesses to faint and fall into another witness.)

Edward L. Harper, executed on a spring day in Kentucky in 1999, looked peaceful as the three chemicals were pumped into his arm. But autopsy evidence showed it had taken twelve minutes for him to die and that there was a 70 to 100 percent chance that he had been conscious. Death row lawyers claim that he was "tortured to death."[28]

The earliest lethal injection protocol, in Oklahoma, had a casual origin, based on advice solicited by a state senator from a professor at the state's medical school. Other states, acting through their corrections departments and individual prison wardens, copied that protocol. The prison warden who developed the lethal injection in Tennessee testified that he had not talked to anyone with a medical or scientific background.

According to Fordham Law professor Deborah Denno, an authority on death penalty and execution methods, the lack of legal guidelines on execution methods has led to numerous painful botches, which violate the Eighth Amendment. There are countless ways to get it wrong: mistakes in mixing the sodium thiopental (stored as a powder); problems with IV tubes; the possibility that the sedative drug may be diluted by prison personnel who intend to use it for substance abuse. According to one estimate, up to 80 percent of executions in Texas have been botched. In most states, there are no dosage guidelines.[29]

"There are forty botched executions that I know of," Brunner tells me. "Because of the Hippocratic oath, neither doctors nor nurses can get involved in an execution. We shouldn't even examine the prisoner in the execution room. No state publishes the name of the executioner. The state pays in cash and doesn't report it."[30]

Virginia, historically one of the most enthusiastic death

penalty states, has a "death row" and a "death house," according to the death penalty opponent Marie Deans of Amnesty International. "Death row is where you go when you still have appeals left. When your appeals run out, they take you to the death house five to fifteen days in advance. Texas has the same plan, even with a sort of 'green room' for visiting press.[31]

"The intent of the death house is to break them so they don't fight. They shave their heads. They watch them. They take their privacy away. Every conversation is recorded, every use of the toilet. It is noted if they cry. After death row and the death house, it is an enormous relief to die. The indignity will stop."

Deans, who works for the Virginia Mitigation Project, is made to stand outside the chamber when an execution takes place. She suspects it's the nurses who inject but she is not certain. "They might be using foreign doctors who don't know what's going on."[32]

"The death penalty makes good entertainment for the governor's friends," says Barbara C. Sproul, a cofounder of Amnesty International USA. "They are sometimes given front-row seats for the event."[33]

Legal challenges to lethal injection triggered a moratorium on executions in most states—basically in all states—when the Supreme Court announced it would hear a case challenging Kentucky's use of lethal injection. In April 2008, the Court upheld the state's method of putting condemned prisoners to death, although the ruling allows some wiggle room for death row appeals in states with a different lethal recipe. The ruling requires death row inmates to prove a risk of severe pain but also that it is substantial when compared to other methods. Executions in Texas, Alabama, and other southern states with large death rows are likely to resume. Other states, such as California, seem likely to do more soul-searching.[34]

BETTER TO BE A PET HAMSTER

It is instructive to compare lethal injection with the treatment of family pets that are euthanized. The American Veterinary Medical Association has condemned the use of the drug pancuronium either as the sole chemical or when used in combination with the usual euthanasia drug. This has not been lost on condemned prisoners and the lawyers representing them, who forced a number of states to postpone executions until the objections to lethal injections are reviewed. "They're saying I'm less than an animal," said Abu-Ali Abdur'Rahman, a death row inmate at Riverbend Maximum Security Institution in Nashville, Tennessee. "The poison they put into our veins needs to be challenged."[35]

According to the Cornell University College of Veterinary Medicine, the euthanasia of animals is compassionate and painless. "Initially, a pet is made as comfortable as possible," states the Pet Loss Support Hotline.[36] Sometimes a mild sedative or tranquilizer is first given if the animal appears anxious or in pain. The euthanasia is usually a barbiturate, an anesthesia drug, which at higher dosages suppresses the cardiovascular and respiratory systems.

Beth Gatti, a Hadley, Massachusetts, veterinarian, uses a "whopping" dose of sodium pentothal to euthanize suffering animals. "There is no suffering. In less than twenty seconds, they stop breathing." Often the family is present, talking to the animal during the procedure. Gatti euthanizes one or two pets per week, and I asked her when she got used to it. "Never," she said.

"There is no brain death with animals," she adds. "We don't put animals on life support. Only vet schools have life-support capabilities, but it is hard to talk people into letting their pets suffer. People will put their relatives on life support, not their pets."[37]

TRAVELS WITH A DEAD PERSON

In this chapter we have seen that lawyers, medics, and ICU doctors have made death a social construct. We write off people as dead when it is convenient to do so. Patients may now go to the ICU to die, not to be saved. Laws about who is dead and who is alive have led only to more confusion.

Brain death has dramatically changed the legal environment surrounding the declaration of death. In the past, declaring a patient dead who was still alive was about the most serious mistake a doctor could make. Today the law protects doctors whether they're using brain-death criteria or cardiopulmonary criteria. There is evidence that one goal of the Harvard ad hoc committee was to throw a protective shield around the medical profession to protect it against legal action. It succeeded.

In a film made by a drug company, committee member William Sweet laughed about the legal problems of brain death and said that the committee had brought in a hospital's legal firm. "They were just too appalled for words," he said, "at the idea of a dead person who was still breathing and whose heart was still beating." Sweet was not deterred, saying that the legal definition of death is *what the doctor says.* His plan was to get doctors to agree to brain death, that there was no need for new legislation.

George J. Annas, a lawyer specializing in health law, commented, "'You're dead when the doctor says you're dead' is legally correct but politically naive." Kansas was the first state to make brain death legal, but only after Kansas doctors refused to use brain-death criteria unless the legislature passed a statute holding them blameless. Other states followed, but there was no national consistency. You might be dead in Nevada and alive in California. In the decade-plus after the Harvard criteria were introduced, when brain death was a legal definition in some states but not in others, one could put a brain-dead patient into an ambulance on the West Coast and drive across the country, and the patient would go from legally alive to dead and back again. Doctors showed that driving one particular route, the patient would be legally alive, dead, alive, dead, dead, alive, alive, dead, dead, alive, alive as the ambulance crossed state lines.[38]

The Uniform Determination of Death Act (UDDA) of 1981 changed all that, but it wasn't really uniform. Yes, declaring a patient dead via brain-death criteria was now legal in every state, but the UDDA was vague, stating "The medical profession remains free to formulate acceptable medical practices." The UDDA gave each state and, more important, doctors, carte blanche to declare people dead using whatever criteria they, as a group, decided upon. In practice, the criteria are all over the map. Some states require two apnea tests, for example, others only one. Some states require a twenty-four-hour waiting period between the sets of brain-death tests, others as few as two hours. As we've seen, some doctors skip entire tests. Many doctors forget, or don't know, to check for conditions that mimic brain death, such as low body temperature, the presence of barbiturates, and many other conditions, the list of which expands by the year. Sweet's original opinion—you're dead when the doctor says you're dead—is now the law of the land.

In the United Kingdom the law is literally different from but in practice identical to that in the United States. The British do not require "whole-brain" death but simply state that there must be irreversible damage to the brain stem before death can be declared. However, even though the UDDA specifies death of the "entire brain," it doesn't require doctors to test the entire brain, so they are off the hook. In practice, U.S. doctors are required to test only the brain stem, the same as the British. It is presumed that medical personnel at least know the "acceptable medical practices" mandated by the UDDA. However, in 1989, eight full years after passage of the UDDA, Stuart Youngner, MD, a professor of bioethics, psychiatry, and cognitive science at Case Western Reserve University School of Medicine, conducted a survey of 195 physicians and nurses in the United States. Only 35 percent identified the established legal and medical criteria for brain death.[39] In other words, 65 percent didn't know what they were doing. This jibes precisely with Bernat's finding (see chapter 3) that 65 percent of brain-death exams are done incorrectly.

WHO OWNS THE BODY?

If doctors are confused, lawyers are more often clueless. Unfortunately, they are in positions of importance. As mentioned, the celebrated Alan Dershowitz told me that one isn't declared dead until an EEG is conducted. Worse, a former deputy general counsel to the Massachusetts Department of Public Health told me that death is "established by two flat EEGs, at least twenty-four hours apart," a criterion abandoned in 1971. This lawyer regularly counseled institutional health care providers wrestling with end-of-life issues, though obviously he didn't know how the "end of life" was determined.[40]

George Annas makes it clear how important being alive is from a legal point of view: if you're alive, you're in; if you're dead, you're out. Brain-death advocates describe grieving families as emotional, irrational, uninformed, bothersome entities who cannot accept the death of a relative and who don't want the respirator closed down. By contrast, Annas tells the story of a hospital, its doctors, and an OPO that could not come to accept that two parents were not going to give up their son to organ donation and delayed the termination of mechanical support for the patient long after death was declared while they worked on the parents.

On a Friday afternoon in April 1980, twenty-year-old Jeffrey Strachan intentionally shot himself in the head. He was taken to New Jersey's John F. Kennedy Memorial Hospital, put on a ventilator, and pronounced brain dead. The ER doctor told Strachan's parents that their son was brain dead and asked them to donate his organs. A few hours later, a neurologist confirmed the brain-death diagnosis and again asked the parents for their son's organs. They could not decide. The neurologist wrote in his notes that the patient was brain dead and added, "Our staff is working with transplant team personnel in this effort. If they get permission to harvest, proceed." A third attempt was made that same evening by the local OPO, the Delaware Valley Transplant Program, to coax Strachan's organs from the parents. Again they resisted, and the OPO said they would have to decide by eleven the next morning or the organs would deteriorate. The following morning the father informed the ICU that he and his wife did not wish to donate their son's organs and wanted him removed from the ventilator. Despite the alleged deadline stated by the OPO, the ICU doctor made a fourth attempt, asking the parents to further think about donation. He also said he couldn't remove their son from mechanical support until the hospital came up with a procedure for doing so, as if this were some bizarre and extraordinary occurrence.

That evening, however, when the father returned to the hospital, he was informed again that his son was dead but the hospital had not yet disconnected him from mechanical support. On Sunday morning the family was told they could not get their son's body before Monday without a court order. The hospital then did a rare thing: it ran EEGs on Strachan on Sunday and Monday. On Monday morning, Strachan was examined by yet another doctor who confirmed that, yes, he was still dead. That afternoon, the parents were informed that the respirator would be disconnected—but only if they would sign a release absolving the hospital and its doctors of any responsibility.[41]

For three days after requesting that their son be disconnected from the ventilator, two heartbroken parents had to see him lying in bed with tubes in his body, his eyes taped shut, with foam in his mouth. Having refused organ donation several times, they were denied their son's body for three days and then had to sign a release. The Strachans sued and were awarded $140,000 from the hospital for withholding their son's body and for not having the proper procedures for turning off a respirator. This verdict was overturned by the Appellate Division, which said the parents deserved no money because they were merely "bystanders." The court also mocked the parents' rights to their son's body. Though the next of kin has a "right to bury the dead," this "quasi right in property" to the body is "somewhat dubious," it stated. In a third legal go-round, the New Jersey Supreme Court found in favor of the Strachans and against the hospital for its callous handling of its dead patient and his parents.[42]

Like many people in support of the brain-death concept, Annas appears obtuse in his interpretation of the case. He sees the problem as a failure in hospital procedure tied to the fact that there were now two kinds of death. He implies that having both brain death and cardiopulmonary death called simply "death" would solve everything.

Annas seems to miss an obvious interpretation: that the hospital, its doctors, and the OPO were blackmailing the Strachans, holding their son's body as ransom in exchange for his organs. "No organs, no body" seems like a more reasonable subtext than "We don't have a form for turning off the ventilator." Though Annas admits that the doctors involved were "extremely aggressive" about organ donation, he praised the OPO for being proper and professional. To this untrained eye, it looks as though all involved, with the exception of the parents, were emotionally out of control, refusing to take no as an answer from the Strachans to the request for their son's organs.

DON'T DIE IN D.C.

The Uniform Anatomical Gift Act was rewritten in 2006 so that the transplant industry can take organs from dead patients despite objections by their families. However, that act has now been revised, but the original version has already been legislated in many states. The 2006 act must be replaced with the revised version, but at this writing the situation is in flux.[43]

As we saw at the beginning of the chapter, the District of Columbia has a ghoulish law that allows doctors there to "preharvest" a dead patient's organs despite the lack of a donor card or his family's consent while the hospital's "head of decedent services" (an actual title at one institution) tries to locate the family and obtain permission. Particularly vulnerable are tourists and visitors, who are common in D.C. Out-of-towners arrive at the ER—because of an accident or stroke, say—then die, and there is no family in the immediate vicinity to stop the doctors from cutting them open.[44]

This is strangely similar to practices during the era of the European anatomy theaters. The subjects of those events were usually executed prisoners and out-of-towners who had the misfortune to die in a town that hosted an anatomy theater. One "decedent director" said that when the family is finally reached, they almost always agree to donate the dead person's organs—once they've learned the surgery has already been completed. In other words, a father hears that his daughter has died, then that she has been sliced apart as if she were an organ donor. Now he is asked if he wants to donate her organs, the body already having been torn apart. Says Michael DeVita, "You have a procedure that most people would refuse, but then, if you do the procedure, most people accept it and decide to donate." Under these circumstances, he notes, one may be unduly influencing the decision to donate, breaking a cardinal rule of voluntary donation.[45] Many OPOs would like to see the D.C. preharvesting law spread across the nation. In the United States we have a system of "expressed volunteerism" that requires consent from donors and their families, but other countries, such as Spain, Belgium, and Austria, practice a "presumed consent" policy in which citizens must list their objection on a national registry to avoid becoming donors. Anyone else is fair game.[46]

For now, Americans have a choice, except perhaps in D.C., where the option is illusory. Roger W. Evans, the Mayo Clinic's head of health services evaluation, implies that nondonors are akin to murderers. "When people refuse to donate, depriving individuals of organs that could save their lives, maybe we should consider that a homicidal act."[47]

Finally, there is an unspoken attitude that lies beneath the "gift of life" and "finding meaning in tragedy" rhetoric that is given to potential donors and their families. Donors are inevitable nonentities. We celebrate the recipients. The donors are faceless. We criticize donors if it turns out that their livers are too diseased for transplant. Why couldn't she have lived a better life to benefit deserving recipients? In the end, there's a significant element of the transplant business that sees harvesting organs as a punishment, something that happens to the culpable or disenfranchised, such as out-of-towners in Washington, D.C. One of the first major efforts to wrangle large numbers of organs was made by Dr. Jack Kevorkian. Before he got involved in euthanasia, he was promoting organ donation by prisoners on death row. In 2003 bills were proposed in both California and New Mexico that would have made organ donation presumed for motorcyclists not wearing helmets who were declared brain dead as the result of a traffic accident. In our society, bikers are considered "bad" and bikers without helmets are considered even "badder." We punish them by taking their organs.[48]

EIGHT

The Moment of Death and the Search for Self

The proof that we don't understand death is we give dead people a pillow.

—Jerry Seinfeld[1]

MICHAEL DEVITA of the University of Pittsburgh recalls making the rounds at a teaching hospital with his interns in tow when he remembered that he had a patient upstairs who was near death. He sent a few of the young doctors "to check on Mr. Smith" in Room 301 and to report back on whether he was dead yet. DeVita continued rounds with the remainder of the interns, but after some time had passed he wondered what had happened to his emissaries of death. Trotting up to Mr. Smith's room, he found them all paging through "The Washington Manual," the traditional handbook given to interns. But there is nothing in the manual that tells new doctors how to determine which patients are alive and which are dead.[2]

The search for the "atom of life," the one distinct element that, if missing or extinguished, would spell the onset of death, reminds one of the search in neuroscience for the engram, or "memory trace," a group of neurons in the brain that account for a particular memory—your First Communion, say, or the "Pam's Dream" episode of *Dallas.* For twenty-five years, Karl Lashley, the director of the Yerkes Laboratories of Primate Biology in Orange Park, Florida, until 1956, attempted to locate the engram in the brains of rats. He trained them to run mazes, then systematically removed chunk after chunk of cortex and retested them on the same maze. Lashley assumed that sooner or later his scalpel would excise the piece of tissue containing the maze knowledge. Nothing like that ever happened. Rats with massive holes in their brains stumbled, staggered, hobbled, but nevertheless negotiated the maze. They didn't do it well, but they did it. Lashley was forced to conclude that the engram didn't reside in any place in particular.[3]

The effort to pinpoint when a person dies has met with similar failure, the difference being that Lashley had the maturity to admit the fruitlessness of his research, while today's experts strut forth confidently and announce that they have found the "engram" of death, sometimes residing in the heart, sometimes in the heart and lungs, and, most common today, in a tiny structure called the brain stem. We have failed but declared victory.

Most of us would agree that King Tut and the other mummified ancient Egyptians are dead, and that you and I are alive. Somewhere in between these two states lies the moment of death. But where is that? The old standby—and not such a bad standard—is the stopping of the heart. But the stopping of a heart is anything but irreversible. We've seen hearts start up again on their own inside the body, outside the body, even in someone else's body. Christiaan Barnard was the first to show us that a heart could stop in one body and be fired up in another. As for brain death, with the mountain of evidence to the contrary, it is comical to consider that this marks the moment of death, though fifty states accept this legal fiction.

PHOTOGRAPHING THE MOMENT OF DEATH

The search for the moment of death continues, though hampered by the considerable legal apparatus that insists that it has already been found. Let us look at a few brave attempts.

Gregory Sorensen, MD, subjected cats to magnetic resonance imaging (MRI) exams at Massachusetts General Hospital in Boston while he killed them with injections of potassium chloride, the same chemical used by Jack Kevorkian and lethal-injection executioners. Sorenson took MRI images while the cats died, watching to see what happened to their brains during the process. He was hoping to photograph the moment of death.

Sorensen says that the idea of "irreversibility" makes the determination of death problematic. What was irreversible, say, twenty years ago, may be routinely reversible today. He cites the example of strokes. Brain damage from stroke that was irreversible and led irrevocably to death in the 1940s was reversible in the 1980s. In 1996 the FDA approved tissue plasminogen activator (tPA), a clot-dissolving agent, for use against stroke. This drug has increased the reversibility of a stroke from an hour after symptoms begin to three hours. In other words, prior to 1996, MRIs of the brains of stroke victims an hour after the onset of symptoms were putative photographs of the moment of death, or at least brain death. Today those images are meaningless. One can take MRIs for another two hours and still not be sure one is photographing death. What about MRI images taken three hours after the onset of stroke? There is no confidence that that will be the end either. It is safe to assume that medical breakthroughs will continue to make "irreversibility" meaningless.

Sorenson says, "We have not yet quantified when brain death is irreversible." He adds, "No one wants to volunteer for the experiments." As for his dead cats, Sorenson said he could see the tissue infarcting, but he could never pinpoint the moment of death.[4] "I'm not sure MRI can prove that someone who is dead (or a mummy) won't come back to life. As a scientist, you simply have to say such events are extraordinarily rare. As a believer, you can say whatever you'd like; I'm a believer, so I do believe that people will live again . . . but I wouldn't try to use MRI to convince you of that position."[5] Sorenson is a nice, friendly guy, and I hope for his sake that God is not a cat lover.

"What's alive and what's dead breaks down when we get above the cellular level," Sorenson says. "Pathologists don't feel comfortable that a brain is dead until the cell walls break down. True cell death is a daylong process."[6]

DO CELLS SUFFER WHEN THEY DIE?

Lawrence Schwartz says cell death is not so simple. Schwartz is a professor of biology at the University of Massachusetts at Amherst. Like Lynn Margulis, he does not see human beings as the ultimate species, most deserving of study by biologists. Humans are at best a footnote in the history of evolution. "We're a failed experiment," he told me. "We haven't given rise to anything. We're a dead end. We don't adapt to our environment. We adapt our environment to us." Schwartz prefers studying invertebrates, which "have so many mechanisms to deal with the environment and more time to evolve and diversify." He added, "Vertebrates are all boring. A narrow-temperature environment." Like Margulis, Schwartz is also underwhelmed by the accomplishments of molecular biology. "Reading the genome doesn't tell us anything about building the animals any more than a box full of auto parts tells us how to make a car."[7]

Cell death is far removed from brain death. As shown, brain death can be declared when only a few brain cells have actually died. Cells in the remainder of the body are alive and kicking. Brain-dead patients being sustained as beating-heart cadavers are still supplying most of their body's cells with blood and thus oxygen, so total cell death is nowhere in sight. Cell death begins in earnest when the heart stops beating and the lungs cease to breathe. No longer being pumped through the body, the blood will drain from the blood vessels at the top of the body and collect in the lower part. The upper body will become pale, the lower body turning much darker, looking bruised. This is livor mortis.[8]

Even at this point, however, most cells are still not dead. After the heart stops, brain cells will die in a few minutes. Muscle cells can hold on for several hours, and skin and bone cells can stay alive for days. Cells switch from aerobic (with oxygen) respiration to anaerobic (without oxygen) when the blood stops circulating. A by-product of anaerobic respiration is lactic acid, which is what makes your arm muscles hurt during arm wrestling or your legs hurt during a hard run. When you are alive, your blood flow clears out the acid, but in a dead person the body stiffens. This is rigor mortis. Rigor mortis usually begins about three hours after the heart stops and lasts thirty-six hours. Eventually all of the cells die. After rigor mortis come initial decay, putrefaction, black putrefaction, and butyric fermentation. We will not go into all of the steps here; see endnotes for more details.[9] Somewhere in these processes—taking as long as a year, depending on the conditions and the weather—is a moment of death. Where that is may be impossible to determine.

Cell death, though, warns Schwartz, is not always a simple matter. We've been talking here about necrosis, the simpler of two kinds of cell death. It's messy. The cell membrane begins to leak, water and calcium rush in, and the cell bursts.[10] There is also apoptosis, a process of programmed cell death in all of our tissues to remove surplus or defective cells. It is estimated that we lose on average about 1 million cells per second to apoptosis (Greek for "falling off"). This type of cell death is inextricably intertwined with and essential to life. Apoptosis is what allows us to have distinct fingers and toes. When we were embryos we had webbed feet and hands. Apoptosis, a kind of cellular suicide, in the tissue between the digits, is what gave us hands capable of playing the guitar or holding a cigarette. "It is a misregulation of cell death," says Schwartz, "that underlies about seventy percent of human disease."[11] Cancer, for example, results from the ability of cancerous cells to evade apoptosis and grow out of control.

More dramatic is the role of apoptosis in the metamorphosis of caterpillars, tadpoles, and other animals. Cell death in the caterpillar is massive. Any layman can see that there is a huge difference between a caterpillar, the larval stage, and the moth or butterfly, the "perfect insect" that emerges after metamorphosis. The caterpillar, says Schwartz, is a "walking gut." It is indeed "very hungry," as the children's book author Eric Carle wrote. A caterpillar increases in weight 10,000-fold in three weeks. The moth, by contrast, is a fucking machine. The gut is no longer needed for food and is now filled with gonad. Some moths don't have mouths. Eating is not a priority. During metamorphosis, huge numbers of muscle cells die within thirty-six hours. The larva is cannibalized, says Schwartz. The legs are mostly retained, but the brain changes dramatically. The nervous system is retained but is rewired. Different neurons are required to send different directions to the muscles. After all, the animal has gone from crawling to flying.

This begs the question: has the caterpillar died and a moth taken over its cells, killed most of them, and converted them to its own purposes? Molecular biology is of little use here. It tells us only that the caterpillar and the moth that metamorphosed from it have identical DNA. Yet who upon seeing them would say that the caterpillar and the moth are the same individual? What if a caterpillar were to commit a crime in front of witnesses but was not apprehended until after it had undergone metamorphosis and flown away from its chrysalis? Could it be convicted in lepidoptera court? Could it be picked out of a lineup? These may seem like facetious questions, and in part they are. But it strikes at our new definitions of death, which rely more on philosophy—when does the "self" leave the body?—than biology.

The "self" is usually defined as consciousness. Lepidoptera metamorphosis challenges the logic of centering a declaration of death on irreversible lack of consciousness. No biologist claims that the caterpillar dies and a moth is born.[12] It's the same animal. Yet most of the caterpillar's cells have died, irreversibly, particularly in the brain, far more than have died in human patients declared brain dead. Does the moth remember its life as a caterpillar? Does it feel as though it's the same individual as it was during prechrysalis days? Where is the "self" of the caterpillar? There is some evidence, though scanty, that some memories may be maintained. Caterpillars were trained to turn to the left or right at the end of a T-maze, the researchers administering electric shocks to the caterpillars to steer them away from the left to the right and vice versa. Following metamorphosis, the adults, the moths, were made to run a similar maze.

There appeared to be evidence of learning—turning right or left, according to their conditioning—that survived from caterpillar to moth. The experiments, conducted at the University of Massachusetts at Amherst, have been repeated throughout the years, but a level of skepticism remains that consciousness survives metamorphosis.[13]

My point is that by today's brain-death standard for humans, a moth would not be considered alive, its "self" having perished in the chrysalis. Even if one acknowledges the right-turn/left-turn memory, it would not be accepted as substantive consciousness by today's medical establishment, which dismisses such traits as reaction to pain, penile erection, and successful pregnancy as not adding up to a "self."

As for a moment of death, Schwartz says that with apoptosis, there may be a moment when "the Slinky's been pushed down the stairs, committed to death," but with plain-vanilla cell death, necrosis, there is "not a specific moment, the result of a catastrophic event." I asked him when all the cells in a human body are dead. He said, "That's a philosophical question." Then he asked, "Do cells suffer when they die?"[14]

ASYMPTOTIC DEATH

There must be cases of instant or near-instant death. The guillotine must provide an unquestionable example of brain death. But Bryan Young, a Canadian neurologist who is developing standards for brain death in his country, speculates that the eyes would still be functioning and the skull and brain would watch themselves falling into the bucket after the blade severed the neck. UCLA's Alan Shewmon assumes that the people vaporized at ground zero by the nuclear bombs dropped at Hiroshima and Nagasaki experienced a very singular moment of death.

But even those examples are challenged by Linda L. Emanuel. While an internist at the Division of Medical Ethics at Harvard Medical School, she put forth her "asymptotic model of death."[15] Now at the Feinberg School of Medicine at Northwestern University, Emanuel takes issue with the traditional Western understanding of life and death as a dichotomy, rejecting the notion that there is a definable threshold between the two. Her argument is partially semantic, claiming that there is "no *state* of death [emphasis hers]. No one can *be* in a state of death. Once life is lost the individual is *not* and therefore cannot *be* in any state. To say 'she is dead' is meaningless since 'she is' is not compatible with 'dead.'" Emanuel concludes that there is only dying, but no death.

"Asymptote" is a term in geometry. Think of a straight horizontal line stretching toward infinity. Now think of a curving line that approaches the straight line, getting ever closer but never quite reaching the straight line as the lines approach infinity. The straight line is called the asymptote of the curved line. In Emanuel's theory, death is an asymptote, never quite reached by a living organism. In her model, the curved line, as it is about to reach the asymptote, suddenly curves up and away from it. I call it, less poetically, the "pushing up daisies" theory. As organisms die, they set free nutrients for new life. A dead body makes possible the daisies in the earth above it. Emanuel is more scientific, writing about a dying woman: "Early in the process some of Janet's parts would have been capable of integrated function in another organism; later her tissues could still have been harvested successfully for cell culture; and subcellular systems could have been salvaged for in vitro functioning perhaps later still." Because "fragments of biological life" are reused either in natural recycling processes or by transplantation, Emanuel writes, "the totality of biological life does not reach zero, but takes off again into the life curve of other organisms." She even takes issue with victims of massive explosions, saying there's still a "process of dying, albeit extremely contracted." She writes that "the person's tissues or molecules, in some form, can potentially be retrieved."[16] I am not making this up.

Emanuel recommends discarding the double standards of brain and cardiopulmonary death and constructing a zone that includes three different standards, what she calls a "zone of life cessation." Anyone who strays into this zone can be declared dead. At the bottom end, the deadest part of the dead zone, is plain-vanilla death, cardiorespiratory death. In the middle is brain death. And the threshold of the zone, which I would call "death lite," would be neocortical death, higher-brain death. By neocortical death she means patients in persistent vegetative state (PVS), who have a working brain stem, but whose neocortex is theoretically dead.

Scientifically, the asymptotic model tells us little about death; it simply expands the elusive moment of death into a zone, but a zone with definite boundaries. Instead of a point of death, we have two points of death: PVS at the beginning of the zone and cardiopulmonary death at the other. Emanuel eliminates the word "death" with the phrase "life cessation." It's difficult to imagine this catching on with anyone except bureaucrats. Can you imagine this conversation?

"Is Aunt Minnie dead yet?"
"No, but she's in the life cessation zone."

It would be easy to dismiss the asymptotic theory except for two factors: (1) Beyond the semantics and logical contradictions (there is no one moment of death, but there are *two* fine lines that bound a zone of death) lies an ambitious agenda.

Emanuel is describing a huge range of conditions that doctors and society may treat as the equivalents of death (though she doesn't call it "death"). Lumped among those suffering from "life cessation" are PVS patients, many of whom, as we've seen, are sometimes conscious and, in some cases, communicative. It is possible that Emanuel was unaware of the findings that PVS patients may live rich inner lives, which is perhaps just as disturbing as if she were a callous but knowledgeable theorist hoping to unplug them. (Emanuel did not make herself available for an interview on her theory.) Even the late neurologist and brain-death advocate Julius Korein did not like lumping PVS patients in with the dead. As he told *Omni* writer Kathleen Stein, "To consider a vegetative state as death is not practical. If you pronounce them dead and they're breathing on their own, what do you do? Take them out and shoot them? Smother them?"[17]

(2) The asymptotic theory is important because it emerged from Harvard Medical School. Papers with "Harvard" on them, no matter how poorly thought out or lacking in data, tend to have an impact. Recall that the ruminations of thirteen men of the Ad Hoc Committee of the Harvard Medical School to Examine the Determination of Brain Death are now the law in all fifty states.

Emanuel raises the concept of "personhood," the new buzzword among those who wish to determine who is alive and who is dead. She says that "what constitutes personhood is controversial," but it dies slowly, just like the body.

WHAT MADE HER *HER*?
THE ARTIFICE OF PERSONHOOD

"Personhood" is a word that doctors throw around today as if it were a scientific term. Alan Shewmon believes it has nothing to do with medicine but is rather a moral concept. In 2000, at the Third International Symposium on Coma and Death, held in Havana, Shewmon presented evidence that some brain-dead patients are still alive, including a video of a patient who, at the time, had been brain dead for thirteen years. (He would die via cardiopulmonary criteria seven years later.)[18] Despite the fact that this boy, who was on a respirator, passed all brain-death criteria, his shoulder twitched, he sprouted goose bumps, and his hand went into spasms when his arm was lifted by the wrist. Like other brain-dead patients, he healed from his wounds while supposedly dead, and he continued to grow. Gary Greenberg, a writer who covered the symposium for *The New Yorker,* reported that no one took issue with Shewmon's science, but doctors continued to say that brain death was valid because the "person" was missing from such bodies. Most revealing was Fred Plum, the neurologist responsible for the term "persistent vegetative state," who immediately challenged Shewmon at the end of his presentation: "This is anti-Darwinism. The brain is the person, the evolved person, not the machine person. Consciousness is the ultimate. We are not one living cell. We are the evolution of a very large group of systems into the awareness of self and the environment, and that is the production of the civilization in which any of us lives."[19]

There are so many things going on in Plum's statement that it is hard to decide where to begin. The implication here is that we humans are more complicated than other life-forms, that we have consciousness (and they don't, or not very much), and thus that we have a higher bar to hurdle to maintain our personhood. Therefore we need a lower standard of death than chimps or amoebas do.

But is this really Darwinism? I asked Janet Browne, a Harvard historian of science and the preeminent biographer of

Charles Darwin. To Plum's statement she responded, "This is not really anything that Darwin would have given any thought to. However, he certainly believed that there were close connecting links between the animal and human and that humans were not special." I am somewhat glad that Plum is dead, and doesn't have to read Browne's statement. She continued, "Yet in the *Descent of Man* he [Darwin] also agrees with most of his readers that humans have a more developed brain and culture than animals. This is not an absolute difference, merely one of degree."[20] Darwin's name today is invoked almost as often as God's on topics that neither would have given any thought to.

Plum claims that "when the cognitive brain has departed, the person has departed."[21] This might be a common belief among current MDs, but the location of the seat of "personhood" or "soul" or "the self" or whatever has been debated without closure for thousands of years. And there is ample evidence that the "person" is not contained wholly in the brain.

Candace Pert, the discoverer of the opiate receptor in the brain, says that there has been a new paradigm in neuroscience since about 1995. More than three hundred common molecules, chemicals, are found in the brain, the immune system, and the bone marrow. In other words, brain chemicals partly responsible for consciousness are being found all over the body. When Pert says, "The body-mind is one," she's speaking not as a Buddhist but as a biochemist, though Buddhism, she says, may have anticipated this discovery. "Consciousness," she adds, "is a property of the entire body."[22] Just do this mind experiment: Take the brain of Michael Jordan and put it into the body of Woody Allen. Or vice versa. Do you think Jordan's "personhood" would have turned out the same if contained in Allen's body? It's an old mystery, and we will not solve it here.

The brain-death revolution has been driven by organ transplantation but justified in part by the modern form of evolution known as neo-Darwinism (see chapter 1). Lynn Margulis pointed out that neo-Darwinism borrows heavily from creationism for its structure. The creationists see life-forms organized into a tree or pyramid, with humans at the top, and God presiding over it all. Neo-Darwinists have a similar tree with humans at the top and all other life-forms below, with natural selection sitting in for God. Margulis envisioned life in a horizontal structure, more of a bush than a tree, with no species "more evolved" than another. "There is no crown of creation," she said. Browne agrees: "[Darwin] differed from Lamarck, or even from his evolutionary grandfather Dr. Erasmus Darwin, in that he eschewed any 'doctrine of necessary progression' and inner striving towards perfection. . . . He always regarded the chief difference between [Lamarck and him] to be that he, Darwin, did not allow his organisms any future goal, any teleology pulling them forwards, or any internal force that might drive the adaptive changes in specific directions. On the contrary, Darwin's scheme of evolutionary adaptation was based entirely on contingency. Organisms shifted randomly."[23]

The idea of a hierarchical natural world, deterministic and purposeful, with man at the top, is essential to our system of separate death criteria for humans as opposed to all other animals. Margulis called it anthropocentric. "Other animals do not have language," Margulis whimsically pointed out, "and are therefore inferior because they cannot speak of their superiority."[24] The belief that life is rigorously and logically organized and that natural selection proceeds in a rational matter also seems at odds with the original Darwinism of Darwin. Cell-death expert Schwartz says, "There is no grand plan." In fact, if you look at the way organisms work and are structured, he says, "You wouldn't do it this way. It's a Rube Goldberg design."[25]

Shewmon, though not opposing the discontinuation of life support for severely disabled patients, points out that Plum's statement about the self illustrates that doctors are not making medical judgments but rather moral judgments about who deserves to live or die. Let me apologize to doctors before I write what I am about to write. All of my doctors have been terrific, most of them nerds, and many clueless about all things human. They have spent the first part of their lives studying endlessly, the second half working hard to learn their craft and pay back their student loans. Recently, while being prepped for a procedure, I heard the doctor in charge wrapping up a phone call. He was clearly divorced and explaining to someone why he couldn't take his kids this weekend. *Good,* I thought. *A man who neglects his family is more focused on his job.*

But are doctors the people we want to be in charge of determining "personhood"?

In 1989 my father suffered a heart attack followed by a stroke. We are not a sentimental bunch; my dad and I spoke on the phone once a year on his birthday. I flew from Massachusetts to Minneapolis to say good-bye. I found him in the ICU. Upon seeing me, he propped himself up, shook my hand, and said, "Thank you for coming." We had a short talk. It wasn't of Oscar Wilde quality, but it qualified as human-to-human conversation, I thought. He wanted to know if I liked Massachusetts. I had just moved there from Manhattan. A young doctor interrupted us, and I left the room. The doctor approached me later across the hall, and said, "I'm sorry, but your father's not going to regain consciousness before he dies." I mentioned, guardedly, that I had just talked to him. The doctor shook his head sadly. "No. You thought you talked to him. He probably made some reflexive sounds, and you interpreted it as speech." He said not to worry, that many family members delude themselves in this manner. I felt embarrassed, and I believed him. As a science writer I had followed the "talking chimp" controversy of the 1970s, the researchers of Washoe, Koko, et al. fooling themselves into believing that they were having meaningful conversations with their primate subjects. Those scientists were smarter than I. I could certainly, out of wishful thinking, believe that my father had spoken to me when he was just making unconscious, meaningless sounds.

I saw a second, older doctor, my father's internist, enter the room. My father opened his eyes, propped himself up again, shook the man's hand. I was several yards away, out in the hallway, but the two men appeared to be having an animated conversation. When the doctor exited, I said, "It looked like you were talking to him." The doctor said, "Sure. I talk to Cliff every day." I explained what the previous doctor had told me and pointed out the man, who was still on the floor. "Oh," said the older doctor, "that's the neurology resident. They teach them that in medical school today. Everybody is dead or in a coma." A nurse told me that at first my father tried to talk to the resident, but he found him too stupid. The resident would ask what year it was, my father pointing out to the nurse afterward that he was standing next to a calendar when he asked the question. My father would always reply, "Nineteen twenty-nine!" The resident would then ask who the president was. My dad would say, "Herbert Hoover!"[26] Then the resident would shake his head sadly. After a few days of this, my father just pretended to be unconscious when the resident appeared. My dad was ready to die, and at seventy-five, his organs were not coveted. But if you are younger, it's a bad idea to mess with the neurology resident. Your liver could end up in a cooler on its way to Riyadh faster than you can say "Herbert Hoover."

SOME FINAL WORDS ABOUT ORGAN TRANSPLANTS

A common question I've gotten over the past several years is "Why are you campaigning against organ transplantation?" or, more commonly, "Stop saying those things about brain death because it hurts people on the waiting list for organs." As a journalist, my job is not to care one way or the other. I'm simply supposed to provide accurate information. When I was a sportswriter, the home team on occasion had the unhappiness of losing. I nevertheless had to report it, disappoint my readers, and embarrass the coaches. It's the same deal here. A doctor told me (screamed at me, actually) that it was irresponsible of me to write about brain death because it would discourage donors and I would therefore in effect be killing patients on the organ waiting lists. Social engineering is not really the job of the journalist. It is to report facts without regard to the consequences.

Frankly, the transplant industry has little to fear. A $20-billion-per-year business, organ transplantation is well ensconced in U.S. medical and political arenas. A major figure in Barack Obama's White House, for example, favors "presumed consent" when it comes to organ donation. Cass Sunstein, the administrator of the Office of Information and Regulatory Affairs, advocates a policy in which those who die may have their organs harvested unless they have legally indicated their refusal. A possible compromise is "mandated choice." A state government could offer two boxes on a driver's license, and the applicant would be forced to choose organ donation or refuse it in order to get a license.[27] The progression would go from mandated choice to presumed choice, with the final goal being, in my opinion, compulsory organ donation for many Americans. In 2010, there were an estimated 28,144 transplant operations in the United States, and there's plenty of room for growth. Some 111,530 transplant candidates were on waiting lists as of June 2011.[28] About 7,000 waitlist patients per year will die before obtaining their desired organs. We don't know how many years someone else's organs would have extended these patients' lives. What we do know is that they would have contributed $5 billion to the transplant business. No businessman wants to lose $5 billion when he had customers ready to write checks for that amount but could not supply the product.[29] Meanwhile, headlines continue to decry the falling number of traffic fatalities that could fill this demand.

The George W. Bush administration also put its imprimatur on brain death. *Controversies in the Determination of Death,* the December 2008 white paper by the President's Council on Bioethics, thoroughly documented many of the scientific and logical problems with brain death. The eighteen-member council, in its 144-page report, took issue with the two tenets of brain death: (1) that the body of a brain-dead patient "is no longer a 'somatically integrated whole' "; (2) that the patient's blood circulation will stop within a definite span of time. The council concluded, "Both of these supposed facts have been persuasively called into question in recent years." Then, a paragraph later, citing no apparent data, the council found a way to defend brain death, using what it called a "novel and philosophically convincing" argument—namely that a brain-dead patient is "no longer able to carry out the fundamental work of a living organism."[30] A living organism, it decided, must exhibit a " 'needful openness' to the world," which brain-dead patients do not.[31] The council did not declare what clinical tests are required to determine "needful openness" or lack thereof. Presumably ice water and Q-tips would be sufficient.

There appears to be no support from either side of the political spectrum for the organ donor, only the recipient. Sunstein, an Obama appointee and animal lover—he hopes to give animals legal standing to sue for abuse and cruelty—wants everyone to be a presumed donor. Equally enthusiastic about organ transplantation is the former Republican Senate majority leader and admitted cat killer Bill Frist, who is a transplant surgeon. (He confessed to having talked animal shelters into giving him cats as pets and then experimenting on them for projects in medical school—Harvard, of course.) Margaret Lock has noted that in the United States, unlike in Japan, recipients of organs are lionized while the donors have slipped out of the news columns.

There are concerns for others, too. In its report, the President's Council on Bioethics opened the door to using PVS patients and anencephalic newborns. PVS patients, as we've seen, have working brain stems and breathe on their own but are putatively—according to the medical establishment—unconscious. Recent research (see chapter 5), however, illustrates that many PVS patients are conscious enough to communicate with the outside world. Anencephalic babies have intact brain stems, breathe on their own, but are missing most of their cortex. Such donors could be gurneyed to harvest without ventilators. The council showed sensitivity toward surgeons—but not donors—when it advised, "Organ retrieval in such cases might entail the administration of sedatives to the allegedly 'person-less' patient because some signs of continued 'biological life' (such as the open eyes and spontaneous breathing of the PVS patient) would be distracting and disturbing to the surgeons who procure the patient's organs."[32] Shewmon pointed out that there would be the additional distraction of PVS patients screaming during organ retrieval.

Donors and prospective donors appear to have no allies. Liberals and conservatives, the religious Left and Right, all seem to want the organs at any cost. The unborn, fetuses, have plenty of political clout. No one speaks for donors. The press, including me, have traditionally been supportive of the transplant business. The Harvard ethicist and anesthesiologist Robert Truog notes that when *The Boston Globe* and *The New York Times* ran negative stories about organ donation and *60 Minutes* aired a segment about donation after cardiac arrest, "the medical community braced for a resounding backlash from the public." But, he reported, there was essentially no response.[33] Truog's concern here is not the public but the transplant community, which he feels has been too reluctant to violate the dead-donor rule (DDR) for fear of offending the public.[34]

A survey of 1,351 residents in Ohio asked respondents to decide whether three different kinds of patients were dead: the brain dead (86 percent said yes), coma patients (57 percent), and PVS patients (34 percent). More interesting, when asked if they were willing to harvest the organs of the patients they personally considered alive, 45 percent said yes. Truog feels that doctors can expand the donor pool into the untapped ranks of PVS patients without fear of public outrage.[35]

We are a country influenced heavily by economics, which may partly explain the lowly status of the organ donor. The journalist Scott Carney, in his recent book *The Red Market,* told how residents of Tsunami Nagar, a refugee camp in India's Tamil Nadu province for survivors of the 2004 tsunami that hit Indonesia, India, and Sri Lanka, are paid only $800 for their kidneys.[36] Still, that's $800 more than a U.S. donor will receive for all of his organs plus his corneas, skin, long bones, and whatever else the harvest surgical team finds usable.

Paying donors for organs is illegal in the United States. However, Alex Tabarrok, an economist at George Mason University, says, "Being on a transplant team is very lucrative." A heart-lung transplant, for instance, costs $1.1 million, $150,000 going for "procurement" (finding the donor and digging out his heart and lungs).[37] There are lots of fingers in that million-dollar pie. But, says Tabarrok, "the only person not making money is the essential person, the one giving the organs." Who would want to pay all the members of the transplant team if there were no heart and lungs? "The value of the organ," says Tabarrok, "is captured by those who take it. [Its value] is reflected in the price." Tabarrok, by the way, avoids the traditional terms "organ shortage" and "need for organs," favoring "demand for organs."[38]

INFORMED CONSENT?

If you are considering becoming a donor, I suggest you read chapters 3 and 4. As you can see from my correspondence with a local organ procurement organization (in the endnotes for chapter 4), there is no attempt by OPOs to provide informed consent. Truog has been a longtime opponent of brain death and, at the same time, a longtime advocate of organ transplantation, having worked harvests as an anesthesiologist. He says that there is no such thing as informed consent for an organ donor. Normally a surgeon would go through a proposed surgical operation with the patient, explaining the process. Then the patient would have the choice of going ahead with or rejecting the operation. "Checking off a box on your driver's license would not meet the standards of informed consent in any country in the world."[39]

Shewmon agrees. If he volunteers to be a donor, "I would like to know how you are going to know if I'm dead and if you're going to perform an apnea test and how you are going to know that it doesn't make me dead." Shewmon believes that the apnea test is so violently stressful that it can be a self-fulfilling prophecy. He also noted that the alleged separation between doctors who declare brain death and those who retrieve the organs is a flimsy one. "Hospitals are working to increase organ transplants. The OPOs have made amazing inroads into the ICUs." Shewmon does not oppose organ transplants per se, citing the examples of blood transfusions, bone-marrow transplants, and live-kidney transplants in which bodily material is transferred from one living, consenting human to another. As for other organs, he says, "The act of transplantation should not be the cause of death nor the hastening of death." He disagrees with expanding the pool of donors into the untapped market of living PVS patients. "We'd have to change the laws about homicide."[40]

DIE BEFORE YOU DIE

I've learned not to discuss my research with people, having lost at least two friends, both with transplants. I didn't attack them. They were curious about what I had discovered about brain death, and I made the mistake of relating to them, dispassionately, some details. Both had ferreted out information about their donors, who in both cases turned out to be African-American teenagers. Most likely their parents had been told they could turn tragedy into meaning by giving up their bodies. Thus they prolonged the lives of two middle-aged, middle-class white men, one of whom needed a liver due to an epic drinking problem.

My medical insurance company was not happy with me either. A few years into this book, I was diagnosed as diabetic and received a questionnaire in the mail. The insurance carrier stated that diabetics often suffer from depression and it was worried about me. One of the questions was "Do you think about death?" Yes, I do. "How often?" the company wanted to know. "Yearly? Monthly? Weekly? Daily?" And if daily, how many times per day? I dutifully wrote in, "About 70 times per day."

The next time I saw my internist, she told me the insurer had recommended psychotherapy for my severe depression. I explained to her why I thought about death all day—merely an occupational hazard—and she suggested getting therapy nonetheless. I thought, fine, it might help with the research.

The therapist found me tragically undepressed, and I asked her if she could help me design a new life that would maximize the few years that I had left. After all, one should have a different life strategy at sixty than at twenty. She asked why I thought I was going to die and why I had such a great fear of death. I said, I *am* going to die. It's not a fear; it's a reality. There must be some behavior that could be contraindicated for a man my age but other normally dangerous behavior that takes advantage of the fact that I am risking fewer years at sixty or sixty-five years of age than I was at twenty or twenty-five (such as crimes that carry a life sentence, crushing at age twenty but less so at age sixty-five). Surely psychology must have something to say on the topic. Turns out, according to my therapist, it does not. There was therapy for those with terminal illness, for the bereaved, for the about-to-be-bereaved, for professionals who dealt with terminal patients, and so on, but there was nothing for people who were simply aware that their life would come to a natural end. It would seem to me that this is a large, untapped market. The therapist advised me not to think about death.

Instead, I devised my own plan. I began smoking little cigars (though I have never graduated to inhaling). I spoke with Donald Shopland, then of the National Cancer Institute, aka "Mr. Smoke," who advised me to go ahead. Shopland ghostwrote the first Surgeon General's Report back in 1964, warning of the dangers of cigarettes. He said I was too old for the harmful effects of smoking to ever catch up with me, and because I had not started smoking in my teens, I had avoided the most serious damage.[41] I have repelled every suggestion by my doctor to get a prostate test, even though numerous prostateless friends and relatives are urging me to do so. Men over the age of fifty are queuing up by the thousands to have their prostates removed on the slimmest of evidence. The arithmetic doesn't support the rush to the scalpel. Only 28,000 men die per year from prostate cancer, and at the mean age of eighty, whereas 183,000 men are getting their prostates excised or zapped dead with radiation each year. This means that at least 155,000 men must be treated needlessly in order to save the 28,000. And those 28,000 men would on average have lived longer than most, even left untreated. Yet evangelical prostateless men confront me with some regularity about *my* prostate, and when am I going to get rid of it? Buyer's remorse?

I also stopped wasting time on tedious events, such as weddings, half of which result in failed marriages, their success or failure, in any case, not dependent on my attendance or absence. I do make myself available to people going through divorce, in which case my existential affectations can be of comfort. I did not go to either of my parents' funerals. I meant no disrespect. They were dead, after all, and the eulogies delivered would describe people I hardly recognized. I've done riskier things, too, not to be discussed here as the various statutes of limitations have yet to run out.

Beyond that, what is one to do to avoid the pitfalls not of death per se but of the awkward period leading up to it? This is actually a problem for all people of any age, since death can take us at any time, but younger people have better reasons for denying death. Older folk are tragically unprepared, and the set routines are fairly superficial. For example, I know that I must start shopping for a white belt with matching loafers. I'm trying to skip lunch so I'll be hungry for dinner at 3:30 and thus be able to take advantage of the early-bird special. My wife says I must soon make a decision: either buckle my belt under my protruding stomach or pull my pants up close to my armpits. Beyond that there is little guidance. I asked Skidmore College psychologist Sheldon Solomon, whom we met in chapter 1, to comment on areas in which we might be blowing it, where we might be compromising our lives due to the denial of death. We selected the following topics:

Genealogy. A sudden fascination with one's ancestors can be a sign of death denial. Solomon says that now you can spit into a cup, have your DNA analyzed, and be told if you're related to Thomas Jefferson or Genghis Khan. If you are descended from royalty, that is a sign that you are immortal, as kings were once considered related to God. Or that at least was the original appeal of genealogy.

Memoir. Solomon credits the ancient Chinese writer Ge Hong (284–364), who wanted to live forever, with inventing the autobiographical memoir. The idea, says Solomon, is that "the physical sinks into the earth, but praise lasts forever." Today, vanity presses have emerged that specialize in writing and printing personal and family memoirs for people that no one wants to read about but who are willing to pay handsomely for the illusion that someone does. I have writer friends who have done such work, and they say the key is to write only flattering things about the subject and nothing controversial, which means that such works are doomed to dullness.[42]

Wills and trusts. A will is a way of reaching out from the grave and manipulating the world when one is a corpse. My wife and I contrived a long list of conditions and age benchmarks our son would have to meet before he inherited our estate—that of two freelance writers—in a concatenation of minuscule steps. Our lawyer talked us into a simple will instead. Even so, the cost of the will was an appreciable fraction of the so-called estate. Solomon found that after catastrophic events, such as the terrorist attacks of September 11, 2001, will and estate lawyers do a booming business.

Fame. We think of younger people as being starstruck, but the desire for fame increases with proximity to death, says Solomon. "We wish to insinuate ourselves into the collective neural net." Yet we still die.

A VERY FEW WORDS ABOUT RELIGION

Thinking incessantly about death did at times impose some costs. For a few years I would wake up very frightened, thinking I was being smothered or buried or dying in some fashion. I wanted to be revived or to be fully dead, but not trapped in this limbo state. That was a physical fear, more about dying than death per se. It was replaced by another troubling thought: annihilation. When I die, I will no longer exist. What I consider *myself* will be yanked forever out of the universe. Perhaps you are reading this page in a library in the year 2040 (if libraries exist then). You are of no comfort to me, even if you like the narrator of this book. I am gone, no matter what it says on the copyright page.

I try to grasp this concept, a universe with no me in it, at least once a week. Though it is comforting—such as when a cop is hassling me or an editor says he is deleting my contact information from his computer—I have yet to comprehend it. A world without me in it? How does that work?[43]

This is where religion comes in. Religion and philosophy are probably better tools for understanding death than, as we have seen, medical science, which is big on confidence but low on results. Unfortunately, I do not understand religion. Some good people have tried to teach me, with no luck. I'm just not wired up to receive the information. If you are religious, I envy you.

As for the specifics of some religions, such as God or a hereafter,[44] those are not within my security clearance. I asked Barbara C. Sproul, the director of the religion program at Hunter College, what hope religion, particularly Christianity, holds for giving us another life after death. She said (all emphasis hers):

> People have several reasons for believing in an afterlife, the major one being egocentrism and the consequent discomfort of facing their inevitable non-existence. It also allows them to believe in some sort of relative meaning in life, because of the judgment to follow. One tends to evaluate such beliefs in terms of what's in it for the believers, and in this case the impetus seems to be on the side of what Freud would call an illusion, a belief resulting from wish fulfillment.
>
> Most religions speak of *eternal* life, which by definition isn't *after* anything; it is always now, forever. If you define yourself as an instance of eternal being, a child of God, one in whom Christ lives (in Christianity), then you don't die when you die. You die before when you give up egocentrism, when you live in faith, understanding your being in the context of God. This is what is meant by "sacrificing yourself." A similar point is made in Buddhism where people are encouraged to "die before you die." But if you identify with ego-self, then you die when you die and that's where people read afterlife into the promise of eternal life. I think it is a perversion of the point but surely it is a popular one.
>
> Afterlife *is* referred to in various Biblical and Koranic texts where judgment is discussed and rewards

> are promised. There is a great deal of dualism in most monotheistic religions, even though that isn't their main point—and is in fact contradictory to it. In general, eternal life is what is promised and afterlife is extrapolated, the same way one speaks of God as "almighty" and then talks of infinite power as being *over* us—i.e. as being finite. If God is all powerful, then when I stand up I do so with divine power. One can speak of a future reward meaning after death or after the death of egocentrism when you finally understand your eternal being. But one cannot speak of eternal being in itself as future; that is afterlife—after a mortal, separate, finite (egocentrically understood) life.[45]

If I understand Sproul correctly, should I desire an afterlife, I must bypass actual religion and go to church instead. But if I want eternity, I can have that now, though Sproul says I must sacrifice (literally "to make holy") my egocentric self. Unfortunately, that is the self I like most. Sproul points out that even St. Augustine prayed to be saved, "but not yet."

There are more religious people around than you might think. Carl Sagan, for example, as his death approached, talked about being star dust. Atoms were formed inside hot stars, he said, which exploded into space, and formed the earth and us. He said we are all made of "star stuff," and after we die and turn to dust, the process will be repeated. Sagan, a proud atheist,[46] would be chagrined to hear that this is a religious statement. Says Sproul, "Sagan is defining himself in the context of absolute reality, which is a quintessential religious outlook."

Another atheist who betrays a religious side is the neo-Darwinist Richard Dawkins, who has made a good living ridiculing religious folk. Dawkins is a materialist, a nonbeliever in

God because there is no physical evidence. Yet, as an Episcopal priest pointed out to me, Dawkins dedicated one of his recent books, "To my parents."[47] Dawkins's parents are both dead. They do not exist. In fact, the existence of God can be neither confirmed nor denied via physical evidence, whereas Mr. and Mrs. Dawkins can be physically confirmed as nonexistent. I am not making fun of every person who has dedicated a book to dead parents or done something to memorialize their parents, but this shows that Dawkins's alleged hard-nosed materialism applies only to other people's beliefs, not to his own cherished memories.[48] It reminds me a bit of Robert Truog, who wants anesthetics for his family members but ridicules your concern that your son or daughter might be in pain.

Solomon says there are two kinds of atheists: (1) the benignly agnostic type and (2) the shrill, hysterical type, to which Dawkins and Sagan belong. The second can be tested for death denial just like Christian fundamentalists. In his mortality salience experiments, reminding subjects of their mortality causes them to defend their worldviews ferociously. The converse is also true: challenging their beliefs can remind them of their mortality. When Solomon et al. showed inconsistencies in the Bible to Christian fundamentalists, thoughts of their own mortality emerged. When devout atheists were shown a phony paper in which Harvard scientists stated that they had found evidence that Jesus was real and had actually been resurrected, they thought of death.[49]

Robert Trivers, perhaps this country's leading neo-Darwinist, who wrote the introduction to Dawkins's breakout book, *The Selfish Gene,* nonetheless is annoyed with what he calls his colleague's "evangelical atheism." I was surprised that Trivers was not an atheist. "Then what are you?" I asked. "I'm just scared," he said. Perhaps we could start a new religion, The Church of Just Scared.

CHECKING OUT

One of the most impressive people I met during the time I volunteered for hospice was Norma Palazzo, the spiritual and bereavement counselor at the Fisher Home hospice in Amherst, Massachusetts. She calls herself an "on-call midwife for death." As I also found she says the dying do not go through great transformations in their final days. "We die the way we live," says Palazzo. She is less interested in the NDE than the NDA, for near-death awareness. Some predictable things happen: a surge of energy, picking at the air with the fingers.[50]

Family and friends are often in as much, or more, pain than the dying patient. One husband leapt atop his dying wife, shaking her. "How can you do this to me? This isn't supposed to happen." To a grieving father, Palazzo might say, "What are you going to be able to take from your daughter after she's gone?" One woman, as she was dying, said, "It's so goddamn beautiful." Her husband was standing by her bed, holding her dancing shoes.

Religious beliefs can help the dying. A man with a previous NDE was not afraid of death, but his wife didn't want him to die. He in fact was looking forward to death. Palazzo has seen some slight conversions among the survivors. "Atheists will become Unitarians." She has heard many regrets. "People regret what they didn't do. They don't regret things they did wrong." She thought a moment. "Unless they're Catholic. Then it's 'Get the priest in here.'"

In his way, Ronald Lashway seems as spiritually centered as Palazzo. Remarkably serene but with a dry wit, Lashway is the funeral director and owner of the Douglass Funeral Service in my town. One spring day he gave me and my hospice colleagues a tour of his facilities.[51] He showed us the nerve center, the embalming room with its scary tubs and hoses. The state of Massachusetts requires only two years of embalming training, but Lashway spent an additional twelve years as an apprentice to learn his craft. The embalming fluid is now six times as strong in concentration as when he began in the business because of the high doses of antibiotics people take. They destroy the blood vessels, and a higher concentration of embalming fluid is needed to make up for the seepage. He sets the facial features first, before draining the blood. After that the body becomes dehydrated and the features are set. The worst bodies to work on are organ donors. They have been savaged, and once some OPOs have their organs, they don't care about returning the bodies to the families in a timely manner. "You can't ever tell the family when the funeral might happen," he said.

It takes a full day to cremate a body, and then they still have to take out the long bones and grind them up. Lashway goes out of his way to keep death a secret. Unlike in the TV series *Six Feet Under,* he never picks up a body in a hearse. He sends a van so as to be inconspicuous, although the license plates are hearse plates. The nice thing about the mortuary business is that there are predictable seasons for death. "People die in numbers before and after major holidays but rarely *on* those holidays. We usually get Thanksgiving, Christmas, and Easter off, so it's very convenient." He speculates that people with terminal illness have some control over the timing of their deaths, either wanting to celebrate a final Christmas, say, or dying beforehand to avoid spoiling their family's holiday.

Lashway has probably dealt with far more dead bodies than the average MD. I asked him what his thirty-five years of embalming people had taught him about death. He answered, "Nothing. Death remains a mystery."

Lashway was by far the most modest person I'd met in all

of my interviewing. When it comes to death, it is unknowable, and we are all on our own. Death, by definition, is irreversible, so there is no one to tell us what it is, despite all the self-assured statements I've heard from medical scientists and others who see death on a regular basis. You must depend on your instincts and your own common sense. "Dying alone" appears to be redundant. It's the only way it's done. For some people it can still be dignified and beautiful.

I did not meet the most remarkable patient to use the hospice. My director told me the story of a woman we'll call Emily. She had called earlier in the day to ask if a room was available. There was. Later, the director heard a knock on the door. There was Emily with only a walking stick and a small satchel of belongings. Most patients are brought by family or arrive in an ambulence or van. Emily had taken the bus. The paperwork is very complicated and normally takes days, with the aid of friends and family, to complete. Medical insurance, state and federal aid, disposition of the body, belongings, pets, and so on. Emily was fully prepared, and she and the director finished all the paperwork in forty-five minutes, a record. Emily then asked to see her room. The director obliged and asked if she wanted to talk awhile. Emily said no, she wanted to take a nap. The director said she'd wake her in a couple hours for dinner.

Dinner came, and the director entered the room to awaken Emily. She did not respond, so the director felt Emily's pulse. She was gone.

Notes

One: Death Is Here to Stay

1. The President's Council on Bioethics, *Controversies in the Determination of Death,* December, 2008, p. 73. Specifically, the council wrote that some dead people have healthier organs than others, depending on the method used to declare them dead.

2. Early Darwinism promoted an organism's instinct to survive. The more modern approach, espoused by W. D. Hamilton and other neo-Darwinists, is that natural selection favors attributes that promote the perpetuation of genes rather than self-preservation per se. Still, the two are generally intertwined.

3. Becker's evidence that other animals are unaware of their impending death is very weak.

4. Ernest Becker, *The Denial of Death* (New York: Free Press, 1973), p. 17.

5. Sam Keen, in the foreword to Becker, *The Denial of Death.* Keen, a *Psychology Today* editor, interviewed Becker as he lay dying in the hospital, refusing painkillers so he could remain coherent. Keen summarized Becker's ideas in the foreword and was more concise than Becker himself.

6. Sheldon Solomon, quoted in "Essential Science Indicators (ESI)," October 2005, p. 2.

7. Keen, foreword to Becker, *The Denial of Death,* p. xiii.

8. Becker, *The Denial of Death,* p. 27.

9. Telephone interview with Sheldon Solomon, August 11, 2006. When "death" was flashed on a computer screen, control subjects saw instead a subliminal "field" for the same length of time.

10. Ibid.

11. Sheldon Solomon, quoted in "Essential Science Indicators (ESI)," October 2005, p. 3.

12. Telephone interview with Sheldon Solomon, August 11, 2006.

13. Florette Cohen, Sheldon Solomon, Molly Maxfield, Tom Pyszczynski, and Jeff Greenberg, "Fatal Attraction: The Effects of Mortality Salience on Evaluations of Charismatic, Task-Oriented, and Relationship-Oriented Leaders," *Psychological Science* 15, no. 12 (2004): 846–851.

14. Lea Winerman, "The Politics of Mortality," *APA Monitor* 36, no. 1 (January 2005): 2–4.

15. In all honesty, not all of these people died on my watch, which was very short, but none made it to age sixty.

16. U.S. Department of Transportation.

17. Kathy Keeton, who died of breast cancer at age fifty-eight.

18. The figures and information in this section on "how many people have ever lived" come from a variety of sources. The pioneer on this topic is the demographer Nathan Keyfitz. See his book *Applied Mathematical Demography* (New York: John Wiley and Sons, 1976). Carl Haub, of the Population Reference Bureau, summed up the work on the topic in his article "How Many People Have Ever Lived on Earth?" *Population Today,* February 1995. Ronald Lee, a professor of demography and economics at the University of California at Berkeley, interpreted the data in a phone interview with the author in July 1996.

19. The modesty of demographers is admirable in that they admit that their estimates are questionable. They are far more scientific than cosmologists, who tell us, with unwavering certitude, what happened at the beginning of the universe some 10 billion years or so ago.

20. Children would have been a liability among hunter-gatherer societies, and scholars suspect that infanticide was a common means of population control.

21. N. R. Kleinfield, "Diabetes and Its Awful Toll Quietly Emerge as a Crisis," *The New York Times,* January 9, 2006. I pick on *The New York Times* not because it is a bad newspaper but because it is most likely the best. To be fair to the *Times,* I don't doubt that the facts in the series are accurate, just that they are not placed in a proper context. I call it "atypically overwrought" because the *Times* is famous for its restraint. This piece is not restrained.

22. Ibid.

23. Varian C. Brandon of the National Center for Health Marketing, Centers for Disease Control and Prevention in Atlanta, Georgia, provided these statistics, which vary from figures supplied by the American Diabetes Association, which claims that diabetes is the fifth deadliest disease.

24. Sherwin B. Nuland, *How We Die* (New York: Knopf, 1993), p. 72.

25. Ibid., p. 79.

26. Ibid.

27. Ibid., pp. 80–81.

28. Ibid., p. 83.

29. Stephen S. Hall, *Merchants of Immortality* (Boston: Houghton Mifflin, 2003), pp. 26–27. To put it more precisely, it's not that biology had nothing to say about aging. Hall wrote that if the orthodox view were correct, cellular biology had nothing to say. But it was not correct, as Hall very quickly establishes.

30. Ibid., pp. 25–26.

31. Ibid., p. 29.

32. Ibid., p. 43.

33. Comfort is probably better known as the author of *The Joy of Sex.*

34. Full disclosure: Hall reviewed my previous book, *Lost Discoveries,* in *The New York Times Book Review.*

35. Conversations and e-mails between Sargent and author over a several-month period.

36. During a lunch conversation on January 21, 1994, in Amherst, Mass. I did not contact Guccione with this request. I hadn't worked for him in more than five years at that point. I don't know that I would have anyway. The models for *Penthouse,* to my knowledge, are not stockpiled in any one place.

37. Drake Bennett, "The Evolutionary Revolutionary," *The Boston Globe,* March 27, 2005. This is a very good article and summary of Trivers's life, with good pictures of Trivers. Trivers also has little use for Richard Dawkins, whose breakout book, *The Selfish Gene,* is based on the work of W. D. Hamilton. Hamilton said we must take a "gene's eye view" of evolution, what the gene wants rather than what the life form wants. By changing this to "the selfish gene," Dawkins anthropomorphized the gene and at the same time borrowed the concept from Hamilton. About Dawkins, Trivers told me, "He's one of those scientists who lays claim to a field by renaming it."

38. The one kingdom that doesn't die is bacteria. You can kill bacteria, but they do not self-destruct in a programmed way as we humans do.

39. Dawkins's *The Selfish Gene* and Wilson's *Sociobiology* borrow significantly from Hamilton's and Trivers's work.

40. The science historian Charles C. Mann, though a staunch evolutionist, points out that neo-Darwinism has a "heads-I-win-tails-you-lose" bent. When a species develops optimally and successfully, natural selection is given credit as a perfect engine of evolution. When it doesn't, that is also proof of natural selection. For example, why do small parts of human bodies fail, rendering them useless and/or dead, when everything else remains functioning well? Hamilton gave the example of car companies' once attempting to improve their product by studying car parts in junkyards, determining which components were still running well. The strategy was to quit trying to improve such parts, as there was no need to; they outlive the car as a whole. (Hamilton never presented evidence that automakers actually carried out such a plan.) But then Hamilton said that such a plan is not rational, because the end result would be that automobiles would fail all at once. All the components would fail simultaneously as we drove down the road. Same deal with the human body: we'd be walking along doing fine and then disintegrate. Thank God natural selection doesn't make perfect bodies! Heads I win, tails you lose.

41. We will not get into explaining what creatures belong in which of the five families. Margulis said that one could boil it down to four kingdoms; the fungi really belong with the protoctists. But scientists, she explained, need to know which meetings to attend, and the fungi people already had their own organizations and meetings in place.

42. Lynn Margulis and Dorion Sagan. *What Is Life?* (New York: Simon & Schuster, 1995), p. 48.

43. Interview with Lynn Margulis, May 24, 2008. The quotes in this paragraph come specifically from the May 24 session, but I interviewed her many times from May 2008 to January 2010 in her home and in her lab at UMass.

44. Sagan was Lynn Margulis's first husband. They had two children together.

45. In all fairness, two other writers were also featured on the cover who had written about the Mars mission for that issue: the venerable Isaac Asimov and the fiction writer Ray Bradbury, the author of *The Martian Chronicles.* They might have boosted sales as much as Sagan. They were the only writers cited on the cover, which carried just one cover line, "We Land on Mars," with a painting of the Viking Lander on the Martian surface.

46. The original Hail Mary pass was a last-second, desperation, come-from-behind, fifty-yard, game-winning pass from Dallas Cowboys quarterback Roger Staubach to receiver Drew Pearson in an NFL 1975 divisional playoff game against the Minnesota Vikings. Staubach said, "I closed my eyes and said a Hail Mary." The expression now refers to any desperate act that has little chance of success.

47. For editors Jeffrey Frank and Tina Brown at *The New Yorker,* both now long gone. Thanks to J. Frank for letting me out of my contract.

48. I was one of only two men to fail the physical that day. The other guy was almost blind. I was rejected for having inexplicably swollen fingers.

49. Telephone interview with Candace Pert, June 10, 2011.

Two: A History of Death

1. Jessica Snyder, *Corpse.* (Cambridge, Mass.: Perseus Publishing, 2001), p. 128. The entomologist was Jerry Payne, whose work with insects that eat dead pigs (used because they are hairless and thus not unlike humans) furthered criminal forensics.

2. Interview with Sue Audette, assistant town clerk, Amherst, Mass., October 12, 2009.

3. Figures supplied by Sandy (she wouldn't give her last name) at the CDC.

4. J. Worth Estes, *The Medical Skills of Ancient Egypt* (Science History Publications, 1989), p. 78.

5. Louis T. Kircos and Emily Teeter, "Studying the Mummy of Petosiris:

A Preliminary Report," *The Oriental Institute News and Notes,* September–October 1991, p. 6.

6. Estes, *The Medical Skills of Ancient Egypt,* p. 40.

7. Ibid., p. 41.

8. Telephone conversation with Peter Lacovara, June 24, 1998.

9. John Baines and Peter Lacovara, "Death, the Dead, and Burial in Ancient Egyptian Society," paper delivered at the American Research Center in Egypt, New York, 1996, pp. 16–21.

10. Jacques-Benigne Winslow, *The Uncertainty of the Signs of Death and the Danger of Precipitate Interments and Dissections, Demonstrated.* Trans. J.-J. Bruhier d'Ablaincourt (London: printed for M. Cooper, 1746). From electronic reproduction by Thomson Gale, Farmington Hills, Mich., 2003, of the original in the British Library, pp. 110–112.

11. Telephone conversation with Peter Lacovara, June 24, 1998.

12. Ibid.

13. Ibid.

14. Telephone interview with Emily Teeter, June 24, 1998.

15. Mary M. Farrell and Daniel L. Levin, "Brain Death in the Pediatric Patient: Historical, Sociological, Medical, Religious, Cultural, Legal, and Ethical Considerations," *Critical Care Medicine* 21, no. 12 (1993): 1953.

16. Hippocrates, trans. Cornarius, quoted in Nancy G. Siraisi, *The Clock and the Mirror: Girolamo Cardano and Renaissance Medicine* (Princeton, N.J.: Princeton University Press, 1997), p. 67.

17. Farrell and Levin, "Brain Death in the Pediatric Patient," p. 1953.

18. W. R. Albury, "Ideas of Life and Death," in *Companion Encyclopedia of the History of Medicine,* vol. 1, ed. W. F. Bynum and Roy Porter (London: Routledge, 1993), pp. 249–251.

19. Paul Carrick. *Medical Ethics in Antiquity* (Dordrecht: D. Reidel Publishing Co., c. 1985), pp. 52, 53.

20. Farrell and Levin, "Brain Death in the Pediatric Patient," p. 1952.

21. Ibid., p. 1953.

22. William Tebb and Edward Perry Vollum, *Premature Burial and How It May Be Prevented* (London: Swan Sonnenscheim, 1905), p. 12.

23. Plato, quoted in Dilberus, *Tom. I, Disput. Philol,* quoted by d'Ablaincourt in Winslow, *The Uncertainty of the Signs of Death,* p. 93.

24. *Plato's Republic,* Book 10, quoted in Winslow, pp. 28, 29.

25. Johannes Andreas Quenstedt, quoted in *Korman's Treatise de Miraculis Mortuorum,* quoted in Winslow, pp. 29, 30.

26. Aulus Cornelius Celsus, quoted in Winslow, pp. 80, 81.

27. Tebb, *Premature Burial,* p. 12.

28. Marcus Fabius Quintilian, quoted in Lancisi, *Tr. de Mort. Subit.* (*De subitaneis mortibus* [1707; "On Sudden Death"] L. I. Cap. 15), quoted in Winslow, p. 27.

29. Quoted in Winslow, p. 122.

30. Alfred C. Rush, *Death and Burial in Christian Antiquity* (Washington, D.C.: Catholic University of America Press, 1941), p. 108.

31. Avraham Steinberg, trans. Fred Rosner, *Encyclopedia of Jewish Medical Ethics: A Compilation of Jewish Medical Law* (Jerusalem: Feldheim Publishers, c. 2003), p. 699.

32. Fred Rosner, trans. and ed., *Julius Preuss' Biblical and Talmudic Medicine* (New York: Hebrew Publishing Co., 1978), pp. 187, 188.

33. Ibid., p. 511.

34. Ibid., p. 512.

35. Psalms 56:7, quoted in ibid., p. 237.

36. Ibid., p. 237.

37. Ibid., pp. 235, 236.

38. Responsa Shevet Halevi and Responsa Maha, quoted in Steinberg, *Encyclopedia of Jewish Medical Ethics,* p. 699.

39. Steinberg, p. 701.

40. Quenstedt, quoted in Winslow, p. 116.

41. Darrel W. Amundsen, *Medicine, Society and Faith in the Ancient and Medieval Worlds* (Baltimore/London: John Hopkins University Press, 1996), pp. 74, 75.

42. Ibid., pp. 76, 77.

43. Bilinda Straight, *Miracles and Extraordinary Experience in Northern Kenya* (Philadelphia: University of Pennsylvania, 2007), pp. 138–141.

44. Rush, pp. 91, 92.

45. Ibid., pp. 101–103.

46. Farrell and Levin, "Brain Death in the Pediatric Patient," pp. 1951–1952.

47. Vivian Nutton, "Medicine in Medieval Western Europe, 1000–1500," in Lawrence Conrad, Michael Neve, Vivian Nutton, Roy Porter, and Andrew Wear, *The Western Medical Tradition: 800 BC to AD 1800* (Cambridge, England: Cambridge University Press, 1995), p. 175.

48. Ibid., pp. 175, 176.

49. Ibid., p. 176.

50. Straight, *Miracles and Extraordinary Experience in Northern Kenya,* pp. 139, 140.

51. Ibid., p. 140.

52. Ibid., p. 137.

53. Ibid., pp. 135, 136.

54. Pedro Lopez-Gallo. "A Man Came . . . His Name Was John," *The BC Catholic,* January 17, 2000, http://bcc.rcav.org/00-10-17/c-gallo.htm.

55. Greg Tobin, *Selecting the Pope* (New York: Barnes & Noble, 2003), p. 55.

56. There is a saying in journalism that some stories are just too good to check. I made the mistake of checking this one. The camerlengo has a good job with few duties. His tenure runs concurrently with the pope's, but he doesn't spring into action until the pontiff dies. Then, as noted, he stands over the pope's body and calls out his baptismal name three times. Or, as Cardinal Eugène Tisserant, Pope Pius XII's camerlengo, reportedly said thrice, "Eugenio, are you still alive?"

If the pope fails to answer, the camerlengo lightly taps the pontifical forehead three times with a silver hammer. If there is no response, the pope is officially dead. Prior to those tests, the attending physicians have already determined the pope's death, but he's not officially dead until the camerlengo says he is.

I can confirm the above story to the satisfaction of most fact-checking departments in the world. A dishy tale like this is all over the Internet, but beyond that I've found it in several putatively authoritative books. I've interviewed eight experts on Catholic history, of which seven confirm the camerlengo's silver-hammer procedure; the eighth could neither confirm nor deny. There were some small discrepancies. Some historians say the silver-hammer/whisper-the-baptismal-name ritual was last used on Pope Pius XII. Others claim that the custom was continuous from the Middle Ages through Pope John Paul II.

My sources, from institutions such as Seton Hall, Catholic University,

Mount Holyoke College, and elsewhere, are confident about this most colorful of the camerlengo's duties. But I don't quite trust them. As I pressed all of them for facts, they began to back off. Take Monsignor Robert J. Wister, for example, a professor of Church history at Seton Hall. Wister is also CBS News' consultant on Catholicism, especially papal elections.

Vatican documents do indeed state that the camerlengo holds the responsibility for declaring the pope dead, but they say nothing about silver hammers or calling out names. No technique is specified. Wister, as did everyone I talked to, admitted that there was nothing in the apostolic constitution establishing the silver-hammer/three-name procedure but that the ritual was a "practice rather than a protocol," handed down orally from one camerlengo to the next. Still, I could not find anyone who had witnessed the practice firsthand.

The Vatican has not pinned down the date the office of camerlengo was established. Two candidates are 1059 and 1150. The camerlengo was originally the chamberlain, a household treasurer reporting directly to the pope.

After the pope dies, all of the appointed employees of the Vatican are fired, except for the camerlengo, who seals off the papal goods and destroys the "ring of the fisherman," the pope's seal, to prevent counterfeit papal bulls. He destroys the ring, supposedly using the same silver hammer. The camerlengo rules over the interregnum, arranging for the assembly of the College of Cardinals and the election of a new pope.

A number of strange details have been added through the centuries. The camerlengo decides who, when, or whether photographs of the dead pope can be taken and, if so, makes sure that he is attired in his pontifical vestments. This new duty is the result of ghoulish photos of Pius XII taken by his doctor and sold to tabloids, Elvis style. The camerlengo also has his own money. The Vatican mints its own coins with a picture of the pope; during the period of *sede vacante* ("empty chair"), the camerlengo mints his own coins, which carry his seal, a picture of an umbrella. (I'm told you can occasionally buy a camerlengo coin on eBay.)

I also found that strange things have happened to dead popes. For example, though most popes are buried in Saint Peter's Basilica, the internal organs of seventy-five popes have been embalmed separately and stored in a vault at the Cathedral of Saints Vincent and Anastasias, adjacent to the Trevi Fountain (a technique copied from the ancient Egyptians, who took out the innards to help preserve the rest of the body in times before competent embalming). This practice ended after the death of Pope Leo XIII in 1903, after which time popes were embalmed conventionally. Unfortunately, Italians are not good embalmers. John Paul I, on view at Saint

Peter's, had to be moved because of the smell. "His head was green," said Wister, who attended the viewing. Pius XII was displayed in a large plastic bag. He did not look good.

57. Andrew Wear, "Medicine in Early Modern Europe," in Conrad et al., *The Western Medical Tradition,* pp. 241, 242.

58. Ibid., p. 242.

59. Siraisi, *The Clock and the Mirror,* p. 75; O. Cameron Gruner, trans., *A Treatise on the Canon of Medicine of Avicenna,* incorporating a translation of the first book (London: Luzac & Co., 1930), pp. 71, 72.

60. Siraisi, *The Clock and the Mirror,* p. 75.

61. Ibid., pp. 67, 68.

62. Wear, "Medicine in Early Modern Europe," p. 242.

63. Albury, "Ideas of Life and Death," p. 252.

64. Farrell and Levin, "Brain Death in the Pediatric Patient," p. 1954.

65. Straight, *Miracles and Extraordinary Experience in Northern Kenya,* p. 140.

66. William Shakespeare, *King Lear,* Act V, Scene iii, ll. 260–293, in *The Complete Works of Shakespeare,* ed. William Aldis Wright (Garden City, N.Y.: Doubleday, 1936), p. 1023.

67. Quoted in Winslow, *The Uncertainty of the Signs of Death,* pp. 40, 45.

68. Quoted in Winslow, p. 40.

69. Quoted in Winslow, p. 44.

70. Paolo Zacchias, *Tom. III. Conf. 79. N. 21,* quoted in Winslow, p. 31.

71. Terilli, physician in Venice, in *Tr. de Causis Mort. Repentin,* sect. vi, cap. 2, quoted in Winslow, p. 30.

72. The section on anatomy theaters was taken from two sources, heavily conflated here. They are: (1) Katharine Park, "The Criminal and the Saintly Body: Autopsy and Dissection in Renaissance Italy," *Renaissance Quarterly* 47, no. 1. (Spring 1994), pp. 1–33; (2) Giovanna Ferrari, "Public Anatomy Lessons and the Carnival: The Anatomy Theatre of Bologna," *Past and Present* 117, no. 1 (November 1987), pp. 50–106. Thanks to Dava Sobel for finding these articles and mailing them to me.

73. Roy Porter, "The Eighteenth Century," in Conrad et al., *The Western Medical Tradition,* p. 475.

74. Albury, quoted in Bynum and Porter, *Companion Encyclopedia of the History of Medicine,* pp. 252, 253.

75. Porter, "The Eighteenth Century," p. 475.

76. Ibid.

77. Farrell and Levin, "Brain Death in the Pediatric Patient," p. 1953.

78. Ibid.

79. Benjamin Franklin, "Restoration of Life by Sun Rays," letter to Barbeu Dubourg, 1773, in *Scientific Deductions & Conjectures,* pp. 150–152.

80. Farrell and Levin, "Brain Death in the Pediatric Patient," p. 1953.

81. James Blake Bailey, *The Diary of a Resurrectionist, 1811–1812,* to which are added an account of the Resurrection Men in London and a "Short History of the Passing of the Anatomy Act." (London: Swan Sonnenschein, 1896), pp. 13–15.

82. Ibid., pp. 19, 20.

83. Ibid., pp. 48, 49.

84. Giovanni Maria Lancisi, *Tr. de Mort. Subit.,* L. I. cap. 15 (*De subitaneis mortibus:* 1707; "On Sudden Death"), quoted in Winslow, *The Uncertainty of the Signs of Death,* p. 28.

85. Steinberg, *Encyclopedia of Jewish Medical Ethics,* p. 696.

86. Lawrence E. Gibson, "Livedo Reticularis: When Is It a Concern?," www.mayoclinic.com/health/livedo-reticularis/AN01622.

87. Steinberg, *Encyclopedia of Jewish Medical Ethics,* p. 696.

88. Winslow, *The Uncertainty of the Signs of Death.*

89. Ibid., pp. 11–15.

90. Farrell and Levin, "Brain Death in the Pediatric Patient," p. 1954.

91. Winslow, *The Uncertainty of the Signs of Death,* pp. 16–20.

92. Farrell and Levin, "Brain Death in the Pediatric Patient," p. 1954.

93. Winslow, *The Uncertainty of the Signs of Death,* pp. 21, 22.

94. Royal Humane Society, "The Very First Award," www.royalhumanesociety.org.uk/html/award_one.html.

95. Ibid.

96. Winslow, *The Uncertainty of the Signs of Death,* pp. 21, 22.

97. Farrell and Levin, "Brain Death in the Pediatric Patient," p. 1954.

98. Ibid.

99. Winslow, *The Uncertainty of the Signs of Death,* pp. 23–25.

100. Farrell and Levin, "Brain Death in the Pediatric Patient," p. 1954.

101. Ibid.

102. A. P. W. Philip, "On the Nature of Death," *Philosophical Transactions of the Royal Society of London* 124 (1834): 169, www.rstl.royalsocietypublishing.org.

103. Farrell and Levin, "Brain Death in the Pediatric Patient," p. 1954.

104. Stanley Joel Reiser, "The Science of Diagnosis—Diagnostic Technology," in *Companion Encyclopedia of the History of Medicine,* vol. 2, ed. W. F. Bynum and Roy Porter (London: Routledge, 1993), p. 828.

105. Albury, "Ideas of Life and Death," pp. 254–255.

106. Paul Brouardel, *Death and Sudden Death* (1902), quoted in Tebb, *Premature Burial,* p. 72.

107. "'One day, towards evening, I was seized with strange and indescribable quiverings. I saw around my bed, innumerable strange faces; they were bright and visionary, and without bodies. There was light and solemnity, and I tried to move, but could not; I could recollect, with perfectness, but the power of motion had departed. I heard the sound of weeping at my pillow, and the voice of the nurse say, 'He is dead.' I cannot describe what I felt at these words. I exerted my utmost power to stir myself, but I could not move even an eyelid. My father drew his hand over my face and closed my eyelids. . . . For three days a number of friends called to see me. I heard them in low accents speak of what I was, and more than one touched me with his finger. The coffin was then procured, and I was laid in it. I felt the coffin lifted and borne away. I heard and felt it placed in the hearse. . . . I felt myself carried on the shoulders of men; I heard the cords of the coffin moved. I felt it swing as dependent by them. It was lowered and rested upon the bottom of the grave. Dreadful was the effort I then made to exert the power of action, but my whole frame was immovable. The sound of the rattling mould as it covered me, was far more tremendous than thunder. This also ceased, and all was silent. This is death, thought I. . . . In the contemplation of this hideous thought, I heard a low sound in the earth over me, and I fancied that the worms and reptiles were coming. The sound continued to grow louder and nearer. Can it be possible, thought I, that my friends suspect that they have buried me too soon? . . . The sound ceased. They dragged me out of the coffin by the head, and carried me swiftly away . . . and by the interchange of one or two brief sentences, I discovered that I was in the hands of two of those robbers, who live by plundering the grave, and selling the bodies of parents, and children, and friends. Being rudely stripped of my shroud, I was placed naked on a table. In a short time I heard by the bustle in the room that the doctors and students were assembling. When all was ready the Demonstrator took his knife, and pierced my bosom. I felt a dreadful crackling . . . a convulsive shudder instantly followed, and a shriek of horror rose from all present. The ice of death was broken up; my trance was ended. The utmost exertions were made to restore me, and in the course of an hour I was in full possession of all my faculties.' " Broadsheet, quoted in Bailey, *The Diary of a Resurrectionist,* pp. 65–68.

108. Tebb, *Premature Burial,* p. 220.

109. Ibid., p. 221.

110. Ibid., p. 223.

111. Ibid., pp. 222, 223.

112. Ibid., pp. 221, 222.

113. Ibid., pp. 224, 225.

114. Ibid., pp. 225–228.

115. Ibid., pp. 228–231.

116. Review, *Gazette Médicale,* recorded in the *Medical Examiner,* Philadelphia, vi, p. 610, quoted in Tebb, pp. 229, 230.

117. Sir Benjamin Ward Richardson, "The Absolute Signs and Proofs of Death," *Asclepiad,* no. 21, 1889, quoted in Tebb, pp. 231–233.

118. Ibid.

1. Respiratory failure, including absence of visible movements of the chest, absence of the respiratory murmur, absence of evidence of transpiration of water vapor from the lungs by breath.

2. Cardiac failure, including absence of arterial pulsation, of cardiac motion, and of cardiac sounds.

3. Absence of turgescence or filling of the veins on making pressure between them and the heart.

4. Reduction of the temperature of the body below the natural standard.

5. Rigor mortis and muscular collapse.

6. Coagulation of the blood.

7. Absence of signs of rust or oxidation of a bright steel blade, after plunging it deep into the tissues (the needle test of Cloquet and Laborde).

8. Absence of red color in semitransparent parts under the influence of a powerful stream of light (diaphanous test).

9. Absence of muscular contraction under the stimulus of galvanism, of heat, and of puncture (electrical stimulation and other pain inducing actions).

10. Absence of red blush of the skin after subcutaneous injection of ammonia (Montiverdi's test) (death will show a brown spot).

11. Putrefactive decomposition.

119. Farrell and Levine, "Brain Death in the Pediatric Patient," p. 1955.

120. Tebb, *Premature Burial,* p. 234.

121. "Death or Coma?," *British Medical Journal,* October 31, 1885, quoted in Tebb, *Premature Burial,* pp. 238–240.

122. "Signs of Death," *British Medical Journal,* September 28, 1895, quoted in Tebb, *Premature Burial,* pp. 237, 238.

123. Tebb, *Premature Burial,* p. 238.

124. Maretus (Marc Antoine Muret), quoted in Winslow, *The Uncertainty of Signs of Death,* p. 113.

125. Quoted in Winslow, *The Uncertainty of Signs of Death,* pp. 113–115.

126. J. J. M. de Groot, *The Religious System of China: Its Ancient Forms, Evolution, History and Present Aspect. Manners, Customers and Social Institutions Connected Therewith* (Leyden: E. J. Brill, 1892), pp. 10, 11, 29, 30.

127. Ibid., pp. 27, 28.

128. Ibid., p. 94.

129. Ibid., p. 104.

130. Winslow, *The Uncertainty of Signs of Death,* pp. 109, 110.

131. Straight, *Miracles and Extraordinary Experience in Northern Kenya,* pp. 118, 119.

132. Ibid., pp. 130, 131.

133. R. H. Codrington, *The Melanesians: Studies in Their Anthropology and Folk-Lore* (New Haven, Conn.: Hraf Press, 1957), pp. 254, 255, 261, 263, 266, 267.

134. Ibid., p. 268.

Three: The Brain-Death Revolution

1. Michael A. DeVita, "The Death Watch: Certifying Death Using Cardiac Criteria," *Progress in Transplantation* 11, no.1 (2001): 58–66.

2. Ad Hoc Committee of the Harvard Medical School to Examine the Definition of Brain Death, "A Definition of Irreversible Coma," *Journal of the American Medical Association* 205, no. 6 (1968): 337–338.

3. When I told Charles C. Mann, the author of the bestsellers *1491* and *1493*, and a graduate of Amherst College, that he might one day be declared dead according to a standard called "the Harvard criteria," he said, "I'm scared already."

4. A. Mohandas, and Shelley N. Chou, "Brain Death: A Clinical and Pathological Study," *Journal of Neurosurgery* 35, no. 2 (1971): 211–218.

5. "Falsification" is a concept popularized by the philosopher Karl Popper that for the most part scientists have accepted as a practical method of testing theories. If genetic theory predicts that a man and a woman, both with blue eyes, can have only blue-eyed children and they produce a brown-eyed baby, the theory is wrong. It doesn't matter that the couple might also have seven blue-eyed children, it has been falsified. A good scientist would not say the brown-eyed baby is wrong for having the wrong-color eyes. The Minnesota criteria essentially do that. The series of tests for brain death put forth by the Harvard committee calls for a confirmatory test of an EEG. When two out of nine EEGs demonstrated that the patients were still alive, the Harvard criteria were falsified. What the Minnesota team concluded instead was that the EEGs were wrong. The seven flatline EEGs, on the other hand, were deemed correct.

6. Mohandas and Chou, "Brain Death," p. 217.

7. Ibid.

8. "Three Agents Shot in Drug Buy; 1 Killed, Another Brain Dead," *Tulsa World*, February 6, 1988.

9. "Ockham's Razor," Alan Shewmon interviewed by Natasha Mitchell, February 2, 2003. ABC Radio National.

10. Julius Korein, preface to "Brain Death: Interrelated Medical and Social Issues." The New York Academy of Sciences, New York, N.Y., 1978, p. 2.

11. Ibid., p. 7.

12. Ibid., pp. 63–64.

13. "Defining Death," *Time*, March 10, 1975.

14. Gaetano F. Molinari, "Review of Clinical Criteria of Brain Death," *Annals of the New York Academy of Sciences* 315, New York (November 1978): 62–69, at 65.

15. Ibid., p. 63.

16. Telephone conversation with Michael DeVita, September 4, 1998.

17. Telephone conversation Alan Shewmon, August 10, 1998.

18. Neurosurgeon Peter Black, interviewed in the film "Brain Death: Developing an Understanding," Sandoz Marketing Communications.

19. The section about Dr. Steven Ross is taken from a phone interview on September 16, 1998, plus a September 1998 e-mail from Sharon Clark, public relations associate, Cooper Health System. Dr. Ross's organ tissue affiliation is taken from his bio on recent Cooper University Hospital website.

20. The argument is that the apnea test is so violent that it may in fact kill the patient; see chapter 8.

21. The time period between tests varies from state to state. It is not specified in the UDDA.

22. *University of Pittsburgh Medical Center Policy and Procedure Manual,* January 8, 1997, p. 4.

23. Telephone interview with James Bernat, June 25, 1998.

24. National Conference of Commissioners on Uniform State Laws, Uniform Determination of Death Act. Approved and recommended for enactment in all the states. Approved by the American Medical Association October 19, 1980. Approved by the American Bar Association February 10, 1981.

25. E-mail from Alan Shewmon, September 9, 2011.

26. Gary S. Belkin, "Brain Death and the Historical Understanding of Bioethics," *Journal of the History of Medicine and Allied Sciences* 58, no. 3, (2003): 325–361.

27. President's Council on Bioethics, "Controversies in the Determination of Death: A White Paper by the President's Council on Bioethics," December 2008.

28. Ibid., p. 32.

29. Ibid., p. 34.

30. Molinari, "Clinical Criteria of Brain Death," p. 66.

31. Telephone interview with Alan Shewmon, May 26, 2011.

32. President's Council on Bioethics, "Controversies in the Determination of Death," p. 54.

33. J. L. Bernat, C. M. Culver, and B. Gert, "On the Definition and Criterion of Death," *Annals of Internal Medicine* 94, no. 3 (1981): 391.

34. A. D. Shewmon, "The Brain and Somatic Integration: Insights into the Standard Biological Rationale for Equating 'Brain Death' with Death," *Journal of Medicine and Philosophy* 26, no. 5 (2001): 457–478.

35. President's Council on Bioethics, "Controversies in the Determination of Death," p. 56.

36. E. F. Wijdicks and J. L. Bernat, "Chronic 'Brain Death': Meta-Analysis and Conceptual Consequences," *Neurology* 53, no. 6 (1999): 1538–1545.

37. Telephone interview with Alan Shewmon, May 26, 2011.

38. "Brain Death: Developing an Understanding," film, Sandoz Marketing Communications.

39. T. E. Finucane, Book review: *Brain Death, New England Journal of Medicine,* no. 345 (2002): 786.

40. S. J. Youngner and R. M. Arnold, "Philosophical Debates About the Definition of Death: Who Cares?," *Journal of Medicine and Philosophy* 26, no. 5 (2001): 527–537.

41. J. E. Murray, quoted in Eelco F. M. Wijdicks, "The Neurologist and Harvard Criteria for Brain Death," *Neurology* 61, no. 7 (2003): 970–976.

42. J. E. Murray quoted in ibid.

43. Ibid.

44. Ibid.

45. Ibid.

46. Ibid.

47. Ibid.

48. Francis L. Delmonico, "Interview with Dr. Joseph Murray," *American Journal of Transplantation* 2, no. 9 (2002): 803–806.

49. He is probably referring to patients in persistent vegetative state (PVS). It was once thought that such patients had no higher-brain functions, but that view has changed dramatically in recent years.

50. Ibid.

51. There have been unsuccessful attempts to transplant organs from anencephalic babies, and though the AMA and other factions of the medical community advocate harvesting the organs of those live babies, there has been little public or legislative support for it.

52. Belkin, "Brain Death."

53. P. M. Black and N. T. Zervas, "Declaration of Brain Death in Neurosurgical and Neurological Practice," *Neurosurgery* 15, no. 2 (1984): 170–174.

54. Frank J. Veith, "Brain Death and Organ Transplantation." *Annals of the New York Academy of Sciences* 315 (February 1978): 437. I suspect that the authors would point out that this brain-death determination was done incorrectly (obviously), but Bernat and others have shown this is the norm for brain-death declarations.

55. The patient had "rather minimal hemiparesis," a weakness on one side of the body.

Four: The New Undead

1. "Organ Donor Categories and Management," *University of Pittsburgh Medical Center Policy and Procedure Manual,* p. 1606.

2. Stock Market Watch newsletter, April 2011.

3. *Hastings Center Report* 27, no. 1 (1997): 29.

4. I interviewed McCabe twice, in person at Cooley Dickinson Hospital in Northampton on July 20, 1998, and by telephone on August 11, 1998.

5. Telephone interview with Joanne Lynn, September 4, 1998.

6. "Multiple Organ Recovery for Transplantation," *University of Pittsburgh Medical Center Policy and Procedure Manual,* pp. 8–16.

7. Procedure put together from interview with McCabe combined with ibid.

8. Margaret Lock, *Twice Dead: Organ Transplants and the Reinvention of Death* (Berkeley: University of California Press, 2002).

9. Mary Roach, *Stiff: The Curious Lives of Human Cadavers* (New York: W. W. Norton, 2003).

10. Kathleen Stein, "Last Rights," *Omni,* September 1987.

11. Ibid., p. 60.

12. Roach, *Stiff,* p. 178.

13. Lock, *Twice Dead,* p. 20.

14. Michelle Au, from her website theunderweardrawer.blogspot.com, expanded on in her book *This Won't Hurt a Bit (and Other White Lies): My Education in Medicine and Motherhood* (New York: Grand Central Publishing, 2011).

15. Meeting with intensivists at Baystate Hospital ICU, September 10, 1999.

16. Stein, "Last Rights," p. 60.

17. Gail A. Van Norman. "A Matter of Life and Death: What Every Anesthesiologist Should Know About the Medical, Legal, and Ethical Aspects of Declaring Brain Death," *Anesthesiology* 91, no. 1 (1999): 275–287.

18. Ibid., p. 279.

19. Gregory S. Liptak, "Spontaneous Movements in Brain-Dead Patients—Reply," *The Journal of the American Medical Association* 255, no. 5 (1986): 2028.

20. Lock, *Twice Dead*, p. 116.

21. Annas, interviewed in the film "Brain Death: Developing an Understanding," Sandoz Marketing Communications.

22. Tom E. Woodcock, "New Act Regulating Human Organ Donation Could Facilitate Organ Donation," *British Medical Journal* 324, no. 7345 (2002): 1099.

23. Adrian W. Gelb and Kerri M. Robertson, "Anaesthetic Management of the Brain Dead for Organ Donation," *Canadian Journal of Anaesthesia* 37, no. 7 (2007): 806–812.

24. Telephone conversation with Robert Truog, May 25, 2011.

25. Below, e-mail correspondence with Sean Fitzpatrick of the New England Organ Bank.

> From: Dick Teresi [mailto:dteresi@comcast.net]
>
> Sent: Sunday, October 31, 2010 10:26 PM
>
> To: Community Education
>
> Subject: questions about being a donor
>
> Dear Sirs:
>
> I have some questions before I register as an organ donor. Is there someone who can answer via email?
>
> —dt

On Nov 5, 2010, at 2:07 PM, Sean Fitzpatrick wrote:

Good afternoon, I'd be happy to answer any questions you might have.

Thanks for writing,

Sean

From: Dick Teresi [mailto:dteresi@comcast.net]

Sent: Friday, November 05, 2010 5:13 PM

To: Sean Fitzpatrick

Subject: Re: questions about being a donor

Dear Sean,

Great. Two questions:

1. I assume I will only become a donor if I am declared dead via brain death. Can I be assured that my whole brain will be dead?

2. Can I specify that if I become a donor that I be given anesthetic during the organ retrieval operation? I know this isn't standard, but I'm willing to pay for it. My lawyer says he can set up my estate so that my executor has a fund for paying for such expenses.

—dt

On Nov 8, 2010, at 10:07 AM, Sean Fitzpatrick wrote:

Good morning and thank you for your email.

To be a donor, one must be declared dead by either heart death criteria or brain death criteria. Brain death criteria involves the irreversible cessation of all brain function, including the brain stem. So, yes, if you are an organ donor brain death would include the whole brain. There is no provision, however, for potential organ donors to register their wish to have an anesthetic administered to them after death.

Hope this information helps.

From: Dick Teresi [mailto:dteresi@comcast.net]

Sent: Monday, November 08, 2010 10:26 AM

To: Sean Fitzpatrick

Subject: Re: questions about being a donor

Mr. Fitzpatrick,

Thanks for this. But it raises two further concerns:

1. What is your organization's objection to a donor wishing to have anesthetic administered, especially if the donor makes a firm legal commitment to pay for it? The money could be set aside in escrow from my estate for these purposes.

2. In lieu of that, can I require that, if declared dead via brain-death criteria, an additional test be added to confirm that the neocortex is also shut down?

I realize brain death includes death of the brain stem, but from what I understand there are no routine tests to check whether the neocortex is down, and the cortex is where a person is conscious of pain. I've read a number of articles by anesthesiologists claiming that they've seen pain reactions from donors during the surgical removal of their organs. Obviously, I would like to avoid that. I understand from doctors at Bay State in Springfield, MA, that the diagnostic equipment for testing the cortex is available, but is almost never used. Can I stipulate that a competent test of cortex activity be done before being declared dead? I am willing to do more than just have a donor card, as I realize the family must also approve. I can secure the approval of my next of kin in writing now, while I am still healthy, and make it a condition of my will and their inheritance.

—Dick Teresi

On Nov 8, 2010, at 4:39 PM, Sean Fitzpatrick wrote:

Mr. Teresi,

It is not that NEOB has an objection to having an anesthetic administered, rather it is that given the nature of our organ procurement process it would not be possible for NEOB to promise you now that your legal documents would be presented in a timely way to assure your wishes are followed (note that your death could occur outside of New England where a different organization is responsible for donation). The donation process is time-critical and experience has shown that estate documents are almost never available or accessible to next of kin in the immediate hours after death when a donation decision must be made. It is for this reason that we recommend that individuals who want to be donors register on a donor registry (like that of the Mass. RMV when renewing their drivers license) rather than make a designation in their will. The donor registry database is accessible and may be instantly queried at the time of death. This way a donation decision can be honored within the critically short time frame required. Under current law, no further consent from family is required.

As for the process of brain death declaration, that falls within the role of the hospital and the attending physicians. The system is designed so that there is a clear delineation between death declaration and the donation process to avoid a conflict of interest. No one from the donation and transplant side of the process may be involved in death declaration.

We find that individuals who generally support donation but have specific concerns about the process should be sure to name a health care proxy who knows their wishes and concerns. In that way, should you happen to die in a manner suitable for donation, that proxy will know of your wish to donate but also know of your wishes concerning anesthetic and cortex activity testing.

—Sean

On Nov 10, 2010, at 10:56 AM, Dick Teresi wrote:

Mr. Fitzpatrick,

That's a good idea: naming a health-care proxy. You raise some minor obstacles that are easily hurtled. I can make sure my proxy has the proper legal papers, and in case I should suffer calamity at home unexpectedly, I've checked with my local fire chief, and he suggests also affixing a copy of the papers to my refrigerator addressed to the EMTs. One EMT, he says, will always check the refrigerator door for a DNR order or other such instructions. He suggests keeping a copy in my car as well in the event of an accident. I can make sure other friends and relatives have copies as well. As I understand it, there is time in between the two required sets of tests for brain death to discuss these matters with your organization. This seems preferable to registering as a donor on my driver's license, which would give a hospital carte blanche and the opportunity to ignore my wishes.

It seems to me you're throwing up trivial legal and procedural problems to avoid dealing with the specific issues I've raised. To wit:

1) Does your organization challenge the validity of articles written by anesthesiologists who say organ donors sometimes exhibit all the signs

of pain that living patients undergoing surgery sometimes do when they haven't been given enough anesthetic? Given that I don't relish undergoing a long operation sans anesthetic, I'd like to know if there's any possibility of this so that I may give informed consent.

2) Your organization must know whether the transplant surgeon will allow the anesthesiologist to administer anesthetic in addition to a paralytic. One doctor told me an anesthetic might harm the organs that the transplant team wishes to harvest. Perhaps he is misinformed. I'd like you to clear this up.

Thanks for your trouble.

—Dick Teresi

December 4, 2010

Dear Mr. Fitzpatrick

I haven't heard back from you regarding the email of November 10, below. Perhaps I've overcomplicated it. Can your organization guarantee that I'll have no pain if I'm a donor and I undergo the organ procurement surgical operation? Can you assure me that an independent anesthesiologist observing the procedure will not see evidence of pain responses?

I hope this simplifies my questions.

—Dick Teresi

---

26. In a September 9, 2011, e-mail to the author, Shewmon wrote:

You seem to be requiring empirical science to study or prove something that is intrinsically beyond its domain. By their very nature, subjective, first-person, conscious experiences are not amenable to empirical study. Logical inference is necessarily the best that we can do. Can you prove to me scientifically that stones don't feel pain when you kick them? Can you prove to me that you're not actually a zombie who mimics the external behaviors of a human being and calls himself Dick Teresi?

It is important to understand that neither the absence of an external reaction to noxious stimuli proves absence of a subjective experience of pain, NOR does the presence of physiological stress responses and even reflex withdrawal prove the presence of a subjective experience of pain. If you operate without anesthesia on someone with high cervical quadriplegia, you will get physiological responses just like if you were to operate

> on a neurologically intact person without anesthesia. That's because the intact spinal cord mediates such responses. But if you ask that patient if he/she feels anything, the answer will be no, because the cord cannot relay the pain signals to the brain, which mediates the subjective experience of pain. In the case of brain death, those brain centers are destroyed, not merely disconnected (assuming an accurate diagnosis of brain death—which is another whole issue entirely). Of course you could argue that when in doubt, don't do something that conceivably could cause pain. But you could apply that principle also to prohibitions against picking flowers or kicking stones—since you can't prove to me with evidence that they don't have subjective experiences too. [Emphasis Shewmon's.]

Truog uses the same stone analogy (see page 159).

27. All of Truog's statements in this chapter are from a telephone interview, May 25, 2011.

28. I don't think Truog meant this the way it came out, comparing organ donors to rocks you can kick. He doesn't strike me as an insensitive person, as illustrated by the advice on anesthetics he later gave me.

29. String theorists and cosmologists often claim that their theories are so far ahead of the experimenters' abilities that they cannot be falsified. In fact, I suspect that such thinkers have little interest in testing their theories. Certainly there are ways of falsifying aspects of string theory and cosmology.

30. Telephone interview with Candace Pert, June 10, 2011.

31. In Wallace, *Consider the Lobster and Other Essays* (New York: Little, Brown and Company, 2005), pp. 235–254.

32. Ibid., p. 250.

33. Ibid., p. 248.

34. The late Wallace, a suicide who had suffered from depression, was a terrific writer, but he was primarily a novelist. Having reviewed one of his nonfiction books, I don't think he is quite as good at digging up facts as making them up, and, not that he's lying here, but one wonders about the accuracy of his neurological facts, given that he cites no sources. Still, I quote him because his observations lay bare the wishful thinking of both medical scientists and organizations devoted to lobster consumption.

35. The God and Science Project, headed by Rev. Christopher Carlisle, Amherst, Mass.

36. Telephone conversation with Robert Truog, May 25, 2011.

37. Telephone interview with Lynn Margulis, June 6, 2011.

38. Stein, "Last Rights," p. 116.

39. Lock, *Twice Dead,* p. 22.

40. Natasha Mitchell, "Brain Death," ABC, February 2, 2003.

41. "The Telltale Heart," *Ebony,* March 1968, p. 118.

42. Lock, *Twice Dead,* p. 85.

43. Ibid., p. 86.

44. Hillel A. Shapiro, ed., *Experience with Human Heart Transplantation, Proceedings of the Cape Town Symposium, July 13–16, 1968* (Durban: Butterworths, 1969).

45. M. D. D. Bell, "Early Identification of the Potential Organ Donor: Fundamental Role of Intensive Care or Conflict of Interest?," *Intensive Care Medicine* 36, no. 9 (2010): 1451–1453.

46. John Shea, "The Inconvenient Truth About Organ Donations," *Catholic Insight,* September 2007.

47. Michael A. DeVita and Arthur Caplan, "Caring for Donors or Patients? Ethical Concerns About the Uniform Anatomical Gift Act (2006)," *Annals of Internal Medicine* 147, no. 12 (2008): 876–879.

48. William P. Dillon et al., "Life Support and Maternal Death During Pregnancy," *The Journal of the American Medical Association* 248, no. 9 (1982): 1089–1091.

49. Mark Siegler and Daniel Wikler, "Brain Death and Live Birth," *The Journal of the American Medical Association* 248, no. 9 (1982): 1101–1102.

50. Rachel A. Farragher and John G. Laffey, "Maternal Brain Death and Somatic Support," *Neurocritical Care* 3, no. 2 (2005): 99–106.

51. Anita J. Catlin and Deborah Volat, "When the Fetus Is Alive but the Mother Is Not: Critical Care Somatic Support as an Accepted Model of Care in the Twenty-First Century?," *Critical Care Nursing Clinics of North America* 21, no. 2 (2009): 267–276.

52. Farragher and Laffey, "Maternal Brain Death and Somatic Support," p. 103.

53. Ibid., p. 101.

54. Jerome B. Posner, "Alleged Awakenings from Prolonged Coma and Brain Death and Delivery of Live Babies from Brain-Dead Mothers Do Not Negate Brain Death," *The Signs of Death. The Proceedings of the Working Group. 11–12 September 2006* (Vatican City: Pontificia Academia Scientiarum, 2007), pp. 116–122.

55. Telephone interview with David Haig, July 27, 1998. We had an interesting talk. Haig had helped me earlier for a *New York Times* story on the theoretical possibility of male pregnancy in humans. He thought it was not an insurmountable problem in that the placenta is such an aggressive, clever structure: it doesn't need a womb and could thrive on any blood supply within its host.

Five: Netherworlds

1. "Kate's Story," directed and produced by Zara Hayes (2008), available at http://www.youtube.com/watch?v=A04AvsGH0FU (accessed September 29, 2011).

2. Rebecca Morelle, "I Felt Trapped Inside My Body," September 7, 2006, BBC News, http://news.bbc.co.uk/go/pr/fr/-/2/hi/health/5321460.stm.

3. Hayes, "Kate's Story."

4. Ibid.

5. Ibid.

6. Hayes, "Kate's Story"; Morelle, "I Felt Trapped Inside My Body."

7. In 1983, a presidential commission estimated that there were five thousand vegetative patients in the United States; fewer than ten years later, the American Medical Association put the number at twenty times higher. The Multi-Society Task Force figure is fourteen thousand to thirty-five thousand. Obviously, there is no scrupulous tally of these patients. Peter John McCullagh. *Conscious in a Vegetative State? A Critique of the PVS Concept* (Kluver Academic Publications, 2004).

8. Fred Plum and Jerome B. Posner, *Diagnosis of Stupor and Coma* (F. A. Davis & Company, 1972).

9. The Multi-Society Task Force on PVS, "Medical Aspects of the Persistent Vegetative State," *The New England Journal of Medicine* 330 (1994): 1499–1508.

10. "Persistent Vegetative State and the Decision to Withdraw or Withhold Life Support," Joint Report of the Council on Ethical and Judicial Affairs and the Council on Scientific Affairs of the American Medical Association, June 1989.

11. Ford Vox, "New Research May Help Unlock Vegetative and Minimally Conscious Patients," *Slate,* April 23, 2009.

12. Jean-Dominique Bauby, *The Diving Bell and the Butterfly* (New York: Alfred A. Knopf, 1997). (Originally published as *Le Scaphandre et le papillon* [Paris: Editions Robert Laffont, 1997].)

13. Interview with Richard Melia, associate division director, Sciences Research Division, National Institute on Disability and Rehabilitation Research (NIDRR).

14. Interview with Marinos C. Dalakas, chief, Neuromuscular Diseases Section, Medical Neurology Branch, National Institute of Neurological Disorders and Stroke (NINDS).

15. Shiv Kumar Rajdev, et al., "Guillain Barre Syndrome Mimicking Cerebral Death." *Indian Journal of Critical Care Medicine* 7, no. 1 (2003): 50–52.

16. Chris Borthwick, "The Permanent Vegetative State: Ethical Crux, Medical Fiction?" Available at http://home.vicnet.net.au/%7Eborth/PVSILM.htm (accessed September 29, 2011).

17. Nancy L. Childs, MD, et al., "Accuracy of Diagnosis of Persistent Vegetative State," *Neurology* 43 (1993): 145.

18. Keith Andrews, et al., "Misdiagnosis of the Vegetative State: Retrospective Study in a Rehabilitation Unit," *British Medical Journal* 313, no. 7048 (1996): 13–16.

19. Ronald Cranford, "Misdiagnosing the Persistent Vegetative State," *British Medical Journal* 313, no. 7048 (1996): 5–6.

20. R. E. Cranford, "The Persistent Vegetative State: The Medical Reality (Getting the Facts Straight)," The Hastings Center Report, 1988.

21. Chris Borthwick, "The Proof of the Vegetable: A Commentary on Ethical Futility," *Journal of Medical Ethics* 21, no. 4 (1995): 205–208.

22. National Center for Ethics in Health Care, "Caring for Patients in Persistent Vegetative State (PVS)," issued March 2005.

23. Childs et al., "Accuracy of Diagnosis": "A diagnosis of persistent vegetative state in the U.S. usually still requires the petitioner to prove in court that recovery is impossible by informed medical opinion."

24. K. Mitchell, I. Kerridge, and T. Lovat, "Medical Futility, Treatment Withdrawal, and the Persistent Vegetative State," *Journal of Medical Ethics* 19, no. 2 (1993): 71–76.

25. "Terry Wallis, a Modern Lazarus: Man Wakes After 20 Years in Coma," available at www.everything2.com/index.pl?node_id=1475825.

26. "Man's Brain Rewired Itself in 19 Years After Crash," msnbc.com.

27. Nicholas D. Schiff interviewed by Anderson Cooper, CNN, September 14, 2006.

28. Steven Laureys et al., "Brain Function in the Vegetative State," *Advances in Experimental Medicine and Biology,* no. 550 (2004): 229–238.

29. A. M. Owen et al. "Residual Auditory Function in Persistent Vegetative State: A Combined PET and fMRI Study," *Neuropsychological Rehabilitation* 15, nos. 3–4 (2005): 290–306.

30. A. M. Owen et al., "Detecting Awareness in the Vegetative State," *Science* 313, no. 5792 (2006): 1376–1379.

31. Sara Boseley, "Think Tennis for Yes, Home for No: How Doctors Helped Man in Vegetative State," *The Guardian,* February 3, 2010. Available at www.guardian.co.uk/science/2010/feb/03/vegetative-state-patient-communication (accessed September 29, 2011).

32. Kate Connolly, "Trapped in His Own Body for 23 Years—The Coma Victim Who Screamed Unheard," *The Guardian,* November 23, 2009. Available at www.guardian.co.uk/world/2009/nov/23/man-trapped-coma-23-years (accessed September 29, 2011).

33. M. M. Monti et al., "Willful Modulation of Brain Activity in Disorders of Consciousness," *The New England Journal of Medicine* 362 (2010): 579–589.

34. Celeste Biever, "Steven Laureys: How I Know 'Coma Man' Is Conscious." *New Scientist,* November 27, 2009. The interview includes the following:

> Q. *Did you ever communicate with him in any other way?*
>
> A. He has undergone a very extensive medical and neurological assessment—but as his physician I cannot tell you more. . . . Do you want me to put his medical record on the internet, or show the videos we made for his assessment? I don't think you would like it if I put results of your IQ test on the internet.
>
> Q. *Can you say what makes you so sure he is conscious?*
>
> A. When I first saw Rom three years ago, he had been diagnosed as being in a vegetative state. We used the Coma Recovery Scale, which is a bedside behavioural assessment done in a very standardized way, and which you do repeatedly so as not to miss any signs of consciousness. And he showed minimal signs of consciousness. So we didn't even need

fancy scanning methods to change the diagnosis. Then he had a brain scan—and we saw near-normal brain function.

35. Nicholas A. Schiff, "Understanding the Recovery of Consciousness," lecture at SUNY-Stonybrook sponsored by the Swartz Foundation, March 15, 2010.

36. George Melendez's story was featured in *60 Minutes,* "Awakenings: Return to Life," November 21, 2007.

37. Schiff, "Understanding the Recovery of Consciousness."

38. Don Herbert's story was also featured in *60 Minutes,* "Awakenings: Return to Life."

39. "A Discussion About Brain Stimulation" with Ali Rezai, Joseph Giacino, Nicholas Schiff, and Joseph Fins, *Charlie Rose Show,* August 2, 2007. N. D. Schiff et al., "Behavioural Improvements with Thalamic Stimulation After Severe Traumatic Brain Injury," *Nature* 448 (2007): 600–603.

40. Arnold Mindell, *Coma: Key to Awakening* (Boston and Shaftsbury: Shambhalar, 1989), p. 1.

41. Ibid., p. 5.

42. John F. Stins and Steven Laureys, "Thought Translation, Tennis and Turing Tests in the Vegetative State," *Phenomenology and the Cognitive Sciences* 8 (2009): 361–370.

Six: The Near-Death Experience

1. I'm not the first to use this cartoon as an example. I owe this reference to Carol Zaleski, *The Life of the World to Come* (New York: Oxford University Press, 1996), p. 3.

2. A summary of the sequence described in Raymond A. Moody, *Life After Life: The Investigation of a Phenomenon* (New York: Bantam, 1975).

3. Interview with Kenneth Ring, August 1999.

4. Kenneth Ring, *Life at Death: A Scientific Investigation of the Near-Death Experience* (New York: Coward McCann and Geoghenan, 1980).

5. Telephone interview with Kenneth Ring, August 1999.

6. Bruce Greyson, "The Near-Death Experience Scale: Construction, Reliability, and Validity," *Journal of Nervous and Mental Disease* 171, no. 6 (1983): 369–375.

7. The three NDEs above were excerpted from the IANDS website; see www.iands.org/nde_archives/experiencer_accounts/.

8. Tijn Touber, "Life Goes On," *Ode* 3, no. 10 (December 2005). Available at http://www.odemagazine.com/doc/29/life_goes_on, accessed September 29, 2011. P. Van Lommel, R. Van Wees, V. Myers, and I. Elfferich, "Near-Death Experience in Survivors of Cardiac Arrest: A Prospective Study in the Netherlands," *The Lancet* 358, no. 9298 (2001): 2039–2045.

9. Peter Fenwick, as quoted in *Strange but True? Encounters,* "Life Beyond Death" (originally aired October 11, 1996; available at http://www.youtube.com/watch?v=d76zpmtot-w).

10. S. J. Blackmore, G. Brelstaff, K. Nelson, and T. Troscianko, "Is the Richness of Our Visual World an Illusion? Transsaccadic Memory for Complex Scenes," *Perception* 24, no. 9 (1995): 1075–1081.

11. Susan Blackmore, "There Is No Stream of Consciousness," *Journal of Consciousness Studies* 9, nos. 5–6 (May–June 2002).

12. K. Nelson, M. Mattingly, S. A. Lee, and F. A. Schmitt, "Does the Arousal System Contribute to Near-Death and Out-of-Body Experiences?," *Neurology* 66 (2006): 1003–1009. K. Nelson, M. Mattingly, and F. A. Schmitt, "Out-of-Body Experience and Arousal," *Neurology* 68, no. 10 (2007): 794–795.

13. Jeffrey Long and Janice Miner Holden, "Does the Arousal System Contribute to Near-Death and Out-of-Body Experiences? A Summary and Response," *Journal of Near-Death Studies* 25, no. 3 (2007): 135–169.

14. Interview with Bruce Greyson, June 2, 1995.

15. Touber, "Life Goes On."

16. O. Blanke, S. Ortigue, T. Landis, and M. Seeck, "Stimulating Illusory Own-Body Perceptions," *Nature* 419 (2002): 269–270.

17. Janice M. Holden, Jeffrey Long, and Jason MacLurg, "Out-of-Body Experiences: All in the Brain?," *Journal of Near-Death Experiences* 25 no. 2 (2006): 99–107.

18. Celia Green, *Out-of-the-Body Experiences* (Oxford: Institute of Psychophysical Research, 1968).

19. Interview with Ronald Siegel. Ronald S. Siegel, *Fire in the Brain: Clinical Tales of Hallucination* (New York: Dutton, 1992). An interview with Siegel is quoted in "Are NDEs Hallucinations?," http://www.near-death.com/experiences/lsd/4.html.

20. Interview with Kenneth Ring, April 15, 1995.

21. Karl Jansen, *Ketamine: Dreams and Realities* (Multidisciplinary Association for Psychedelic Studies: 2001) as quoted in http://www.near-death.com/experiences/lsd03.html (accessed September 29, 2011).

22. Karl Jansen, "The Ketamine Model of the Near Death Experience: A Central Role for the NMDA Receptor," *Journal of Near-Death Studies* 16, no. 1 (1997): 5–26, available at http://www.mindspring.com/~scottr/nde/jansen1.html (accessed September 29, 2011).

23. Ibid.

24. Peter Fenwick quoted in John F. Newport, "A Theory That Accounts for the Occurrence of All NDEs," available at http://www.near-death.com/experiences/articles006.html (accessed September 29, 2011).

25. The details of Pam's surgery on pages 279–81, including the quote from Dr. Spetzler, are drawn from Michael B. Sabom, *Light and Death* (Grand Rapids, Mich.: Zondervan Publishing House, 1998), pp. 38–52.

26. The following account is from an interview with "Pam Reynolds."

27. The preceding four paragraphs are from an interview with Michael Sabom, February 10, 2005.

28. Jeffrey Long quoted in "Are NDEs Hallucinations?," available at http://www.near-death.com/experiences/lsd04.html (accessed September 29, 2011).

29. Daniel C. Dennett, *Consciousness Explained* (Boston: Little, Brown and Company, 1991), pp. 5–7.

30. Touber, "Life Goes On."

31. Interview with Michael Sabom, February 10, 1995.

32. Sabom, *Light and Death,* p. 56.

33. Interview with Sabom, February 10, 1995.

34. Jane Dreaper, "Study into Near-Death Experiences," BBC News, September 18, 2008, http://news.bbc.co.uk/2/hi/health/7621608.stm. "World's Largest-Ever Study of Near-Death Experiences," *Science Daily,* September 10, 2008, www.sciencedaily.com/releases/2008/09/080910090829.htm.

35. Interview with Carol Zaleski, January 5, 2005.

36. Interview with Kenneth Ring, June 2, 1998. Distressing aspects of NDEs are discussed in Bruce Greyson and Nancy E. Bush, "Distressing Near-Death Experiences," *Psychiatry* 55, no. 1 (1992): 95.

Seven: Postmodern Death

1. Rick Weiss, "Demand for Organs Fosters Aggressive Collection Methods," *The Washington Post,* November 24, 1997. Mancia died in 1997.

2. Corinne Levy, "Organ Donors: Wanted 'Dead' or 'Alive,'" *The American Journal of Bioethics,* December 20, 2001.

3. Gail A. Van Norman, "Another Matter of Life and Death," *Anesthesiology* 98, no. 3 (2003): 763–773.

4. Norman Paradis, "Are Doctors Taking Organs from Patients Who Are Not Quite Dead?" *60 Minutes,* April 13, 1987.

5. E. H. Albano, "The Medical Examiner's Viewpoint," in *The Moment of Death: A Symposium* (Springfield, Ill.: Charles C. Thomas, 1969).

6. Joseph L. Verheijde, Mohamed Y. Rady, and Joan McGregor, "Presumed Consent for Organ Preservation in Uncontrolled Donation After Cardiac Death in the United States: A Public Policy with Serious Consequences," *Philosophy, Ethics, and Humanities in Medicine,* September 22, 2009, p. 15.

7. Ibid.

8. Meeting with intensivists, Baystate Medical Center, Springfield, Mass., September 10, 1999.

9. Paradis, "Are Surgeons Taking Organs from People Who Are Not Quite Dead?"

10. Michael A. DeVita, "The Death Watch: Certifying Death Using Cardiac Criteria," *Progress in Transplantion* 11, no. 1 (2001): 58–66.

11. Telephone interviews with Michael Baden, June 9, 1999, and June 11, 1999.

12. Jessica Snyder Sachs, *Corpse: Nature, Forensics, and the Struggle to Pinpoint Time of Death* (New York: Perseus Publishing, 2001).

13. Ibid., p. 36.

14. Ibid., p. 42.

15. Richard Selzer quoted in ibid., p. 36.

16. Telephone interview with Alan M. Dershowitz, June 8, 1998.

17. The account of Mel's troubles with the law is based on phone conversation with Alan M. Dershowitz, June 17, 1999, and his book, *The Best Defense* (New York: Vintage Books, 1983).

18. Telephone interview with Barry Scheck, June 8, 1998.

19. Stephanie Barry, "Ex-Officer Who 'Died' Seeks Lively Pay Hike," *The Springfield Union News,* February 11, 2000.

20. Telephone interview with John Neeld, July 12, 1999.

21. Judith E. Nelson and Diane E. Meier, "Palliative Care in the Intensive Care Unit: Part II," *Journal of Intensive Care Medicine* 14, no. 4 (1999): 189–199. T. J. Prendergast et al., "Increasing Incidence of Withholding and Withdrawal of Life Support from the Critically Ill," *American Journal of Respiratory Critical Care Medicine,* no. 155 (1997): 15–20.

22. Nelson and Meier, "Palliative Care."

23. Telephone interview with Robert Kennedy, June 26, 1998.

24. Death Penalty Information Center (www.deathpenaltyinfo.org).

25. Ibid.

26. Telephone interview with John Brunner, July 26, 1995.

27. Adam Liptak, "Critics Say Execution Drug May Hide Suffering," *The New York Times,* October 7, 2003.

28. Adam Liptak, "On Death Row, a Battle Over the Fatal Cocktail," *The New York Times,* September 16, 2004.

29. Telephone interview with Deborah Denno, July 21, 1998.

30. Telephone interview with Brunner, July 26, 1995.

31. Telephone interview with Marie Deans, March 23, 2003.

32. Ibid.

33. Interview with Barbara G. Sproul, August 21, 1998, Bub's Bar-B-Q, Sunderland, Mass.

34. In 2010, thirty-five states and the federal government had death penalty statutes. Twenty offer alternative choices (electrocution, gas, hanging, or firing squad). In 2010 there were seven states with bills to abolish the death penalty; other states have imposed a moratorium; and still others have bills to reinstate capital punishment or expand the applicable crimes.

In 2007 New Jersey became the first state to repeal death penalty laws since capital punishment was reinstated in the United States in 1976. In 2008 the Nebraska Supreme Court ruled that the use of the electric chair violates the state constitution; with no alternative on the books, it is practically without a death penalty.

35. Liptak, "Critics Say Execution Drug May Hide Suffering."

36. Pet Loss Support Hotline, Cornell University College of Veterinary Medicine, http://www.vet.cornell.edu/org/petloss/resources/EuthanasiaSA.htm.

37. Telephone interview with Beth Gatti, August 7, 1998.

38. "Brain Death: Developing an Understanding," film, Sandoz Marketing Communications, was the source for information in this and previous paragraph.

39. Margaret Lock, *Twice Dead: Organ Transplants and the Reinvention of Death* (Berkeley: University of California Press, 2002).

40. E-mail from David C. Harlow, Posternak, Blankstein & Lund, June 19, 1998.

41. George J. Annas, "Brain Death and Organ Donation: You Can Have One Without the Other," *Hastings Center Report* 18, no. 3 (1988): pp. 28–30.

42. The New Jersey Supreme Court did not reinstate the $140,000 award to the Strachans but rather sent the case back for a new trial.

43. Michael A. DeVita and Arthur L. Caplan, "Caring for Organs or for Patients? Ethical Concerns About the Uniform Anatomical Gift Act (2006)," *Annals of Internal Medicine* 147, no. 12 (2008): 876–879.

44. Weiss, "Demand for Organs."

45. Ibid.

46. Kenneth Gundle, "Presumed Consent for Organ Donation," *Stanford Undergraduate Research Journal,* January 4, 2000.

47. Weiss, "Demand for Organs."

48. Stacy Dickert-Conlin, Todd Elder, and Brian Moore, "Donorcycles: Do Motorcycle Helmet Laws Reduce Organ Donations?," Michigan State University, June 10, 2009.

Eight: The Moment of Death and the Search for Self

1. Jerry Seinfeld, *Sein Language* (New York: Bantam Books, 1993), p. 44.

2. Doctors refer to this book, given to interns as a reference guide, simply as "The Washington Manual," but it comes in various editions and under different titles.

3. Judith Hooper and Dick Teresi, *The Three-Pound Universe* (New York: Macmillan, 1986), p. 188.

4. Telephone interview with Gregory Sorensen, July 17, 1998.

5. E-mail from Gregory Sorensen, July 14, 1998.

6. Telephone interview with Sorensen, July 17, 1998.

7. Interview with Lawrence Schwartz, September 16, 2006.

8. Michael Weaver, "What Happens to the Human Body After We Die?," April 25, 2005, www.madsci.org/posts/archives/2005-04/1114460899.Gb.r.html.

9. Ibid. *Initial decay* occurs within three days after the heart stops. Normal bacteria in the intestines begin to feed on the contents of the intestine and the intestine itself. Then they digest other organs. Digestive enzymes leak out and spread through the body, breaking down organs and tissues. Enzymes inside the cells leak out and digest the cell and its connections with other cells. Flies lay their eggs around body openings and wounds, and maggots hatch and feed on the dead tissue. *Putrefaction* occurs four to ten days after the heart stops. The bacteria produce a lot of gas, which attracts more insects. The body inflates, and fluids pour into the body cavity, creating a cozy apartment for the maggots. *Black putrefaction* occurs ten to twenty days after the heart stops. The bloated body now collapses, and the flesh has gotten cottage cheese–like. Exposed parts of the body turn black. The body stinks. Fluids leak out of the body into the soil, attracting more bugs, which will eat most of the flesh. Bacteria also continue to consume the flesh. *Butyric fermentation* begins twenty to fifty days after the heart stops. The flesh has been removed and the body dries out. It has a cheesy smell caused by butyric acid. Mold starts to grow. Beetles are attracted by the smell and replace the maggots, which are no longer able to eat the tough flesh. This stage can take a year. The body is dry and decays slowly; moths and bacteria eat the person's hair, leaving only bones.

10. Elizabeth Luciano, "UMass Biology Professor to Discuss Cell Death," University of Massachusetts press release, March 2, 1999.

11. Interview with Schwartz, September 16, 2006.

12. E-mail from Theodore Sargent, May 16, 2011. Sargent wrote: "I don't think I've ever run into a person who thinks that the caterpillar dies in the metamorphosis from larva to adult. Rather they see it as the caterpillar 'turning into' a moth. Not entirely satisfying, of course, because we could say that the moth 'turns into' an egg."

13. E-mail from Theodore Sargent, professor emeritus, University of Massachusetts, May 16, 2011.

14. Interview with Lawrence Schwartz, September 16, 2006.

15. Linda L. Emanuel, "Reexamining Death: The Asymptotic Model and a Bounded Zone Definition," *Hastings Center Report* 25, no. 4 (1995): 27–35.

16. Ibid., pp. 30–31.

17. Kathleen Stein, "Redefining Death," *Omni,* September 1987, p. 114.

18. Though the boy had met brain-death criteria, under the statutes of the time he was too young to be declared legally dead, and his mother had continued life support.

19. Gary Greenberg, "As Good as Dead," *The New Yorker,* August 13, 2001, pp. 36–41.

20. E-mail from Janet Browne to author, May 17, 2011. Browne is the author of *Charles Darwin: A Biography,* vol. 1, *Voyaging* and *Charles Darwin: A Biography,* vol. 2, *The Power of Place.*

21. Stein, "Redefining Death," p. 114.

22. Telephone interview with Candace Pert, June 10, 2011.

23. Browne, *Charles Darwin: A Biography,* vol. 1, and *Charles Darwin: A Biography,* vol. 2, p. 61.

24. Telephone interview with Lynn Margulis, May 19, 2011.

25. Interview with Schwartz, September 16, 2006.

26. I'd like to point out here that Herbert Hoover was in fact president in 1929, so my father should get at least partial credit for this answer.

27. Matt Cover, "Obama Regulation Czar Advocated Removing People's Organs Without Explicit Consent," September 3, 2009, www.cnsnew.com/node/53534.

28. "2011 U.S. Organ and Tissue Transplant Cost Estimates and Discussion," Milliman Research Report. Milliman is a consulting firm that compiles data for the United Network for Organ Sharing (UNOS).

29. These are back-of-the-envelope calculations on my part. If there were 28,144 transplant operations in 2010, a year in which the transplant industry as a whole grossed $20 billion, then each operation contributed $710,631 to the industry. If we multiply this by the lost 7,000 patients, we get $4.974 billion. Another way to look at it is to examine UNOS's price list, which starts at $259,000 for a kidney and goes up to $1,123,800 for a heart-lung system. Given 7,000 unfulfilled transplant customers per year, this represents a range of lost revenue from $1.8 billion to $7.9 billion.

30. *Controversies in the Determination of Death: A White Paper by the President's Council on Bioethics,* December 2008, p. 90.

31. Ibid., p. 64.

32. Ibid., p. 51.

33. Robert D. Truog, "Brain Death—Too Flawed to Endure, Too Ingrained to Abandon," *The Journal of Law, Medicine & Ethics* 35, no. 2 (2007): 279.

34. Telephone interview with Robert Truog, May 25, 2011.

35. L. A. Siminoff, C. Burant, and S. J. Youngner, "Death and Organ Procurement: Public Beliefs and Attitudes," *Kennedy Institute of Ethics Journal* 14, no. 3 (2004): 217–234.

36. Kate Wong, "Books and Recommendations from *Scientific American*," *Scientific American*, May 27, 2011. A review of Scott Carney, *The Red Market: On the Trail of the World's Organ Brokers, Bone Thieves, Blood Farmers and Child Traffickers* (New York: HarperCollins, 2011).

37. Telephone interview with Alex Tabarrok, November 27, 2010. Figures provided by UNOS. Costs are for 2008 and include only expenses during the first year of the transplant.

38. Telephone interview with Alex Tabarrok, November 27, 2010.

39. Telephone interview with Truog, May 25, 2011.

40. Telephone interview with Shewmon, May 26, 2011.

41. Shopland told me one of tobacco's dirty little secrets. It's starting smoking early that causes "morphological changes" that persist through life, and quitting smoking does little to reverse this damage. He pointed to the suffragette movement, during which women in their thirties began smoking as a symbol of equality with men. There was no incidence of lung cancer among such women.

42. If you desire such a service, contact the Association of Personal Historians, www.personalhistorians.org.

43. Vladimir Nabokov offered the other extreme: being horrified of a universe existing *before* he was born. See *Speak, Memory* (New York: Perigee Books, 1947), p. 19.

44. Not all religions have a god or a hereafter.

45. The e-mail from Barbara Sproul to author, May 30, 2011, finalized her thoughts, but the process went back and forth for many years.

46. Obviously, there are atheists, such as Buddhists, who are religious. Sagan was both an atheist and nonreligious.

47. Thanks to Rev. Christopher Carlisle for bringing this to my attention. The book is *The Blind Watchmaker* by Richard Dawkins.

48. I know, I probably pick on poor Richard Dawkins too much, but it is hard not to. He's such a nasty little pill. I agree with much of what he writes, but he exploits real scientists' work without giving adequate credit until years later. He wraps himself in the mantle of Darwin but appears to ignore his ideas. And it is hard to accept as a Darwinist a man who needed three wives to produce a single offspring. Reproduction is where natural selection really counts. Consider Michael Behe, the intelligent-design proponent and polar opposite of Dawkins. He has nine children. One wonders who is more Darwinian and who will ultimately win the evolutionary struggle. Darwin himself had ten children.

49. Telephone interview with Sheldon Solomon, May 24, 2011.

50. I talked to Norma Palazzo a number of times during my hospice volunteer days, but most of the information here came from a talk she gave to the New Options Group at the Bangs Center in Amherst, Mass., on March 18, 2009.

51. April 23, 2008.

Selected Bibliography

As I began compiling a lengthy bibliography, as I have done for previous books, it occurred to me that I was just bragging about how much work I had done. My sources are adequately cited in the endnotes, and for the interested reader, the notes will lead you to the original material. Below are just a few books, most of them readable, that a reader craving more death material might want to follow up with. I cited books not necessarily because I consider them good but because they are instructive in some manner.

Becker, Ernest. *The Denial of Death.* New York: Free Press, 1973.

Becker's theory—that we all go crazy because of our genetic desire to survive in opposition to the realization that we cannot survive—is way too all-encompassing, in my opinion. But he certainly reveals a core neurosis of human beings.

Cook, Robin. *Coma.* Boston: Little, Brown & Company, 1977.

Sour critics in the death business have criticized Cook for not being careful with his handling of terms (coma, brain death, etc.), but this novel is eerily prescient. Cook writes that his purpose was to highlight a problem that would become more prevalent: the shortage of organs and the lengths to which people would go to get warm bodies. And in the era in which the novel was written, the term "coma" was often used to comprise the whole range of comatose states: brain death, PVS, coma, and so on.

Dennett, Daniel C. *Consciousness Explained.* Boston: Little, Brown and Company, 1991.

Dennett's logic is difficult to follow, but for those who want a glimpse, albeit cloudy, into the neo-Darwinist mind, this book can be useful, if maddening.

Hall, Stephen S. *Merchants of Immortality: Chasing the Dream of Human Life Extension.* Boston and New York: Houghton Mifflin Company, 2003.

Hall reports on the history and cutting edge of life extension science without getting totally sucked in by its devotees' optimism.

Heinrich, Bernd. *Winter World: The Ingenuity of Animal Survival.* New York: HarperCollins, 2003.

Animals destroy many of our notions of what death is, a subject I do not cover. Heinrich writes about how some animals die to survive, and then are resurrected.

Henig, Robin Marantz. "When Does Life Belong to the Living?" *Scientific American,* September 2010, pp. 50–54.

Marantz presents the arguments for and against sacrificing organ donors prior to their deaths to benefit recipients on the waiting list. The article also includes a schematic of the organ harvest of a non-beating-heart cadaver that is reasonably accurate but vastly simplified and sugar-coated.

Korein, Julius, ed. *Brain Death: Interrelated Medical and Social Issues.* New York: The New York Academy of Sciences, 1978.

This is a collection of scientific papers and essays on brain death, but many are comprehensible to the layman. I doubt that many doctors have read this book except for the introduction by Korein, which claims that brain death has been firmly established as real death. Many of Korein's claims have since been falsified, and some of the authors in his very book dispute the reliability of brain death. See especially Gaetano F. Molinari, "Review of Clinical Criteria of Brain Death." Among other sponsors, three drug companies provided "financial assistance," according to the New York Academy of Sciences.

Lock, Margaret. *Twice Dead: Organ Transplants and the Reinvention of Death.* Berkeley: University of California Press, 2002.

By writing extensively about brain death and transplantation in Japan, Lock highlights the differing approach in North America.

Margulis, Lynn, and Dorion Sagan. *What Is Life?* Berkeley: University of California Press, 1995.

A groundbreaking biologist and her coauthor son discuss death, but mostly forms of life—spanning all five kingdoms—that most scientists never consider.

Nuland, Sherwin B. *How We Die: Reflections on Life's Final Chapter.* New York: Alfred A. Knopf, 1994.

A grim, unsentimental, excellent look at what happens to us physically as we die. Nuland, a medical doctor, is a bit of a moralist, however, using words such as "soul," and though he sees death as inevitable, he also considers it to be pathology, describing one patient as "flabby" and "sedentary" and writing of his "gluttony."

Plum, Fred, and Jerome B. Posner. *Diagnosis of Stupor and Coma.* Philadelphia: F. A. Davis, 1972.

I don't actually recommend that anyone read this pioneering book by Plum and Posner, but it's nice to have on one's bookshelf, a great attention getter since few people call in sick with "stupor" anymore. See also the 1812 companion volume, *Cases of Apoplexy and Lethargy.*

Roach, Mary. *Stiff: The Curious Lives of Human Cadavers.* New York: W. W. Norton, 2003.

An entertaining, well-written book by a journalist about the uses of dead people's bodies.

Schrödinger, Erwin. *What Is Life?* London: Cambridge University Press, 1944.

Written by a Nobel Prize–winning physicist—not a doctor or biologist—this is probably the most rigorous, and sensible, attempt to explain life. Being a physicist, Schrödinger is relatively free of the biases and prejudices that cloud the issues around life and death.

Tilney, Nicholas I. *Transplant: From Myth to Reality.* New Haven and London: Yale University Press, 2003.

An interesting history. The author, a Harvard surgeon, emphasizes the need of doctors to get past any "emotional feelings" about the donors and discusses, among other things, schemes to pay African Americans for organ donation because their donation rate is low.

Zaleski, Carol. *The Life of the World to Come.* New York: Oxford University Press, 1996.

Zaleski, a religion scholar at Smith College, discusses various scenarios of life after death, something I don't cover. Zaleski told me that her own religious beliefs pretty much line up with the Catholic catechism.

Index

Page numbers beginning with 293 refer to notes.

About the Author

Dick Teresi is the author or coauthor of several books about science and technology, most recently *The God Particle* and *Lost Discoveries: The Ancient Roots of Modern Science—from the Babylonians to the Maya,* both cited as "notable books" by *The New York Times Book Review.* He is a cofounder of *Omni* magazine and has written for *The New York Times, The Wall Street Journal, Smithsonian, The Atlantic Monthly,* and *Discover,* among others. Teresi has been an editor at eight national magazines, and was editor in chief of four of them: *Omni, Science Digest, Longevity,* and *VQ.*

A Note on the Type

This book was set in Adobe Garamond. Designed for the Adobe Corporation by Robert Slimbach, the fonts are based on types first cut by Claude Garamond (c. 1480–1561). Garamond was a pupil of Geoffroy Tory and is believed to have followed the Venetian models, although he introduced a number of important differences, and it is to him that we owe the letter we now know as "old style." He gave to his letters a certain elegance and feeling of movement that won their creator an immediate reputation and the patronage of Francis I of France.

Typeset by Scribe, Philadelphia, Pennsylvania

Printed and bound by R. R. Donnelley & Sons,
Harrisonburg, Virginia

Designed by Maria Carella